Eduardo Pagel Floriano

Effects of pruning on Pinus elliottii

Eduardo Pagel Floriano

Effects of pruning on Pinus elliottii

Growth as a function of pruning intensity

SciencaScripts

Imprint

Any brand names and product names mentioned in this book are subject to trademark, brand or patent protection and are trademarks or registered trademarks of their respective holders. The use of brand names, product names, common names, trade names, product descriptions etc. even without a particular marking in this work is in no way to be construed to mean that such names may be regarded as unrestricted in respect of trademark and brand protection legislation and could thus be used by anyone.

Cover image: www.ingimage.com

This book is a translation from the original published under ISBN 978-620-6-76050-4.

Publisher:
Sciencia Scripts
is a trademark of
Dodo Books Indian Ocean Ltd. and OmniScriptum S.R.L publishing group

120 High Road, East Finchley, London, N2 9ED, United Kingdom
Str. Armeneasca 28/1, office 1, Chisinau MD-2012, Republic of Moldova, Europe
Printed at: see last page
ISBN: 978-620-7-76062-6

THANKS

Among the many people who, in one way or another, helped directly in the preparation of this work, I must especially mention:

Prof. Dr. Juarez Martins Hoppe (*in memoriam*);

Prof. Dr. César Augusto Guimarães Finger;

Prof. Dr. Paulo Renato Schneider;

Prof. Dr. Sidinei José Lopes;

César Martins Andrade, Forestry Engineer;

Eduardo Righi dos Reis, Forestry Engineer;

Prof. Dr. Magda Lea Bolzan Zanon;

Ftal. Eng. Stefano Leha Dissiuta.

To all of you, and to those who have not been mentioned, sincere thanks for the support and time dedicated to this work and to me personally as a friend.

Eduardo Pagel Floriano

SUMMARY

LIST OF REDUCTIONS

a_i - area of plot of order i (m²).

cm - centimeter.

CV% - coefficient of variation in percentage.

d - diameter of the trunk with bark at breast height (1.3 m height), or DBH.

dsc - diameter of the trunk without bark at breast height (1.3 m height), or DAPsc.

DBH - trunk diameter at breast height (1.3 m high).

MS - average spacing between trees (m).

USA - United States of America.

f - artificial form factor.

F - F statistic.

g - basal area per tree, or cross-sectional area of the trunk at breast level, corresponding to the result of the formula: $g = \pi \cdot d^2 / 4$.

G - basal area per hectare, or sum of the cross-sectional areas of the trunks of all the plants per hectare at breast level, corresponding to the result of: $G = N_{ha} \cdot \pi \cdot d^2 / 4$.

GL - degrees of freedom.

h - total height.

h_{100} - dominant height; height corresponding to the average height of the 100 thickest trees per ha.

ha - hectare.

ICA - Annual current increase.

IU - Indiana University.

m - meter.

m² - square meter.

m³ - cubic meter.

NCSU - North Carolina State University.

N_i - number of trees in the useful area of plot i.

N_{ha} - number of trees per hectare corresponding to the result of the formula: $N_{ha} = 10000.N_i / a_i$.

p. - page.

QM - mean square.

SQ - sum of squares.

t - Student's t-statistic.

r - Pearson's correlation coefficient.

R^2 - coefficient of determination.

$R^2_{aj.}$ - adjusted coefficient of determination.

Reg - regression.

Res - waste.

S% - Relative spacing index as a percentage of the dominant height.

SCP - Secretariat for Coordination and Planning of the State of Rio Grande do Sul.

SEMA-RS - State Secretariat for the Environment of the State of Rio Grande do Sul.

TRAT - Treatment.

UFSM - Federal University of Santa Maria, RS.

USDAFS - United States, Department of Agriculture, Forest Service.

v - total volume with bark per tree.

vcn_{t1+t2} - sum of the average individual volume without bark and with knot of the 1st and 2nd logs.

vsc - average individual volume without bark (m³).

vsn_{t1} - volume without bark and knot of the 1st log.

vsn_{t2} - volume without bark and knots of the 2nd log.

vsn_{t1+t2} - sum of the average individual bark-free and knot-free volume of the 1st and 2nd logs.

W - Shapiro-Wilk W statistic.

χ^2 - Chi-Square statistic.

SUMMARY

EFFECT OF DISRUPTION ON *Pinus elliottii*.

The production of *Pinus* sawnwood accounted for more than 75% of total Brazilian production in 2000 and, among the species of the genus, *P. elliottii* Engelm. has been one of the most widely used in commercial plantations. The quality of the wood of this species is highly influenced by treatments such as pruning, an operation that reduces the productivity of the trees, but is necessary in the production of wood for noble use such as carpentry, considerably increasing its value. With the aim of determining the influence of pruning intensity on the growth and shape of *Pinus elliottii* Engelm., an experiment was set up in a 6-year-old plantation in Piratini, southeastern Rio Grande do Sul, using a randomized block design with four replications and four treatments represented by pruning percentages: 0% (control, no pruning) and up to 40%, 60% and 80% pruning, in relation to the total height of the trees. The operation began when the trees were 6 years old and was corrected annually until the height of 6 meters was reached in all treatments. The results, when the plantation was 15 years old, showed that there was no significant influence of pruning on average height, dominant height and mortality; diameter and volume without bark were significantly influenced by pruning, with the following reductions in relation to the control treatment without pruning: 2.0% of diameter and 3.5% of volume in the 40% pruning treatment; 7.5% of diameter and 15.0% of volume in the 60% pruning treatment; 13.5% of diameter and 26.9% of volume in the 80% pruning intensity treatment. Pruning significantly influenced the shape of the trees, with the pruned treatments having a conical shape with a slight tendency towards paraboloid and the control treatment with no pruning having a conical shape with a slight tendency towards neiloid. The percentage of clean, knot-free wood obtained in the three pruning treatments was similar, reaching an overall average of 46% of the total individual bark-free volume produced up to 15 years of age. The small reduction in performance shown by the 40% pruning intensity treatment, which had no statistically significant influence on production, reinforces the literature on pruning up to this intensity.

1 INTRODUCTION

Historically, all over the world, wood production from planted forests has replaced that from natural forests. A similar phenomenon is occurring in Brazil, with an increase in the production of wood from *Pinus* and other species being a current trend, especially in Rio Grande do Sul, where natural forests have been exploited to the point of complete exhaustion.

The *Pinus elliottii* Engelm. species has been one of the most widely used in commercial plantations in Brazil, mainly for the production of sawn timber for furniture and construction. Wood for these uses implies the need to produce superior quality wood, which can be obtained through silvicultural treatments and appropriate management techniques. *P. elliottii* is also used in the production of softwood pulp, used in the manufacture of packaging; in the production of wood fiber sheets and resin, from which turpentine is extracted.

One of the factors affecting the quality and price of wood is the presence of live or dead knots caused by persistent lateral branches. The natural pruning process is slow and the lateral branches persist in some species, forming knots that are usually dry and loose, which depreciate the wood. More widely spaced stands provide larger timber. However, trees that grow more freely have thicker and more numerous branches. One of the techniques for reducing the size and number of lateral branches is to increase the planting density. However, it has been observed that *P. elliottii* trees growing in dense forest stands have a tendency towards imperfect natural pruning, where the presence of a single branch is enough to depreciate several pieces of wood obtained from a log; in addition, individual growth is reduced due to the high competition between trees, and in the end, smaller logs are obtained, reducing the yield in sawmills and rolling mills.

The method used to prevent the formation of knots in the wood of *Pinus* species is pruning or artificial pruning, a practice that is considered expensive but is generally compensated for by the higher value of the wood produced. However, it has been found that production is influenced by the intensity with which pruning is applied; therefore, determining the maximum intensity with which it should be carried out is of paramount importance for maximizing the production of knot-free wood in *P. elliottii* stands.

In this respect, the aim of this work is to determine, for a *P. elliottii* stand, the influence of pruning intensity on:

- growth in diameter, height and basal area per hectare;
- the shape of the trunk;
- the volume of knot-free wood produced in each pruning treatment.

2 LITERATURE REVIEW

The volume of sawn timber produced in Brazil is over 9 million cubic meters, gradually becoming more represented by the production of planted forests, mainly species of the *Pinus* Linnaeus genus, which accounted for over 75% of Brazilian production in 2000 (Table 1). The total area of planted forests in the state of Rio Grande do Sul corresponds to 480,000 hectares, of which around 180,000 hectares are *Pinus,* contributing more than 700,000 m³ of logs per year (AGEFLOR *apud* Tonini, 2000).

TABLE 1 - Evolution of sawn timber production in Brazil (m³)

REGION	SPECIES	1980	1987	1991	1995	2000
	Araucaria	1.990.000	480.000	300.000	200.000	180.000
SOUTH	Leafy	2.364.000	340.000	210.000	50.000	40.000
	Pine	**130.000**	**1.275.000**	**3.640.000**	**5.000.000**	**7.000.000**
NORTH	Leafy	5.039.000	4.200.000	3.300.000	2.600.000	2.000.000
TOTAL		9.523.000	6.295.000	7.450.000	7.850.000	9.220.000

Source: Revista da Madeira (2000).

According to Persson and Janz (2000), the replacement of native forests by plantations is a historical process that begins with the exploitation of native forests, reducing their natural area of occurrence, part of which is then managed at a percentage that gradually increases until it stabilizes. When around half of the natural areas have already been eliminated and more than 60% of the remaining areas are managed, plantations begin to be carried out which, after a certain period of time, tend to represent around half of the productive forest areas. This process is also taking place in Brazil. As can be seen in Table 1, there has been a gradual increase in the production of planted forests and a reduction in the production of hardwoods, most of which currently come from native forests in Brazil.

Plantations for timber production have been carried out using species with proven productivity, most of which are exotic and among the most widely used are species of the *Pinus* genus, with the *P. elliottii species* already being one of the most widely used in Brazil in the early 1970s (Mattos, n.d.).

According to Gilman and Watson (1994), the species *Pinus elliottii* Engelm. belongs to the Family *Pinaceae,* Order *Coniferales*, Class *Gimnospermae, and* is commonly known as *slash pine* (USA), Florida pine and American pine in Brazil. It is native to southeastern North America and its region of natural occurrence is shown in Figure 1.

According to Gilman and Watson (1994), the species is a large tree with many branches. The bark is gray-brown, dark, heavily furrowed and scaly.

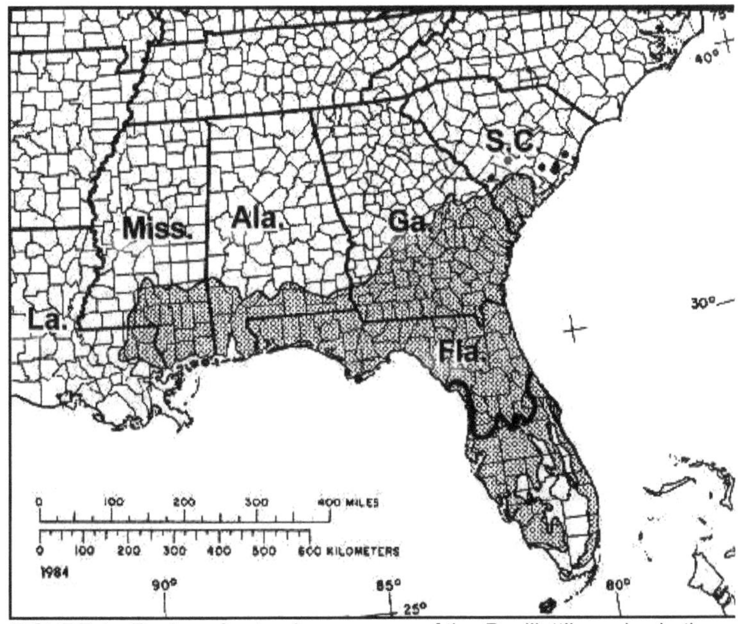

FIGURE 1 - Region of natural occurrence of the *P. elliottii* species in the USA (in gray). Source: NCSU (2004).

It grows in full sun to half shade on clay to sandy, slightly alkaline to acidic, well-drained and occasionally moist soils; it is tolerant of poor soils and moderately tolerant of droughts and saline soils; moist soils with a high pH are harmful to the species. The branches shed naturally and the crown is open, creating light shade; the acicles fall constantly and form a thick blanket below. Its roots are very aggressive, competing intensively with surrounding plants. The seeds spread across the landscape and can become invasive. It is highly susceptible to some diseases and pests that affect its health and aesthetics and can be serious problems, such as chlorosis, wood rot, canker and rust (which oaks have as

13

intermediate hosts); wood borers (beetle larvae) and defoliator fly larvae; among others.

P. elliottii is recommended for windbreaks, shade trees and commercial plantations. Its crown has an irregular silhouette, ranging from oval to pyramidal, with a low density of sparse, medium-thick branches. The foliage is alternate to spiral, the acicles are always green, fragrant, simple and filiform, with entire margins, parallel veins and 10 to 30 cm long. The inflorescences are yellow, inconspicuous and inconspicuous. The female cones are ovo-long, reddish-brown, 8 to 15 cm long by 2.5 to 7.5 cm in diameter, persistent, showy, attractive to rodent mammals. The male strobili are cylindrical, small, 2 to 3 cm long, red to yellow, inserted in groups at the tips of the lateral branches. The trunk and branches tend to grow vertically without drooping, are not showy, have no thorns and are brittle; the wood is also weak and brittle; the new branches (of the year) are thin and brown. The species also produces resin, from which turpentine is made (Clements, 1970). It is fast-growing and grows to around 25-30 m in height, 90-120 cm in diameter at breast height and 10-15 m in crown diameter.

According to Spiecker (1981), tree growth is the result of the effects of their genetic makeup, environmental characteristics (climate, soil, management, etc.) and interactions between the two. From the planting of a forest stand, over time and as biomass, competition and crown proximity increase, mortality also increases and, while growth in trunk and branch diameter is reduced, tree vigor and understory development are also reduced (Oliver *et al.*, 2001). As the trees grow, competition increases and the availability of light, nutrients and water decreases for each individual tree, the lower branches die and, despite the decrease in individual tree growth, the volume stock continues to increase and the radial growth of the trunk is reduced in the lower parts (Kramer and Kozlowski, 1972), resulting in a more cylindrical shape. This makes pruning important, as it eliminates live and dead branches from the trees, providing a greater quantity of wood without knots and, consequently, of better quality and greater commercial value (Gibson *et al.*, 2001), as well as providing greater ease of access and visibility through the forest stand (Warner, 1997).

In the study of the shape of tree trunks, in addition to the artificial shape factor, which is described by Assmann (1970) and Prodan (1997), one of the most widely used functions for studying solids of revolution and associating them with the shape of trees is the function $g_x = p.x^r$, described by Husch *et al.* (1982) and Finger (1992), where the value of "p" is related to the size of the body considered and the value of "r" *is related to the* shape.

In the production of wood for sawmills, wider spacings are used in order to obtain larger timber, but this favors the formation of larger and more numerous branches, resulting in more and larger knots in the wood (Finger *et al.*, 2001).

In addition to causing an undesirable aesthetic appearance, wood knots in most cases negatively affect its physicomechanical quality (Schilling, 1996). Live branches are less depreciating than dead branches and knots from dead branches tend to come loose, creating holes in the sawn or laminated timber (Schneider *et al.*, 1999).

A large part of the tree's volume is found in the first six meters from the ground. Pruning to this height means keeping a large part of the tree's volume free of knots that depreciate the wood (Jeter, 1992).

It has been found that the greater the intensity of artificial pruning, the lower the growth rate of the trees (Kramer and Kozlowski, 1972). Therefore, determining the maximum intensity with which pruning should be carried out is of paramount importance for maximizing the production of knot-free wood in *P. elliottii* stands.

To describe the volume of the trees, Couto and Vettorazzo (1999) tested the fit of the functions for the volume of *P. taeda* : $v = b_0 + b_1 d$, $v = b_0 + b d_1^2$, $v = b + b d_{01}^2 h$, $v = b + b d_{01}^2{}_2 + bh + b_3 d^2 h$, $\ln v = b + b_{01} \ln d$, $\ln v = b + b_{01} \ln d^2 h$ and $\ln v = b_0 + b_1 \ln d + b_2 \ln h$; obtaining low S_{yx} and high r^2 for all the equations, choosing as best the Spurr model $v = b_0 + b_1 d^2 h$ for a 6-year-old stand and the model $v = b_0 + b_1 d^2$ for the same 11-year-old stand. For *P. elliottii* volume estimates, Schneider *et al.* (1999) selected the following equation as the best: $\log v = b + b_{01} \log d + b_2 (\log d)^2 + b_3 \log h + b_4 (\log h)^2$. Where: v=volume (m^3); d=diameter (cm); h=height (m); b , b , b , b_{01234} =parameters; , br^2=coefficient of determination; Syx=standard error of the mean.

Finger *et al.* (2001) tested the following models to study the growth of *Eucalyptus saligna* Smith: $y = b + b_{01} .t$; $y = b + b_{01} .\ln t$; $\ln y = \ln b + b_{01} .\ln t$; $y = b + b_{01} /t$; $y = b + b_{01} .t + b_2 .t^2$; $y = b + b_{01} .t + b .t + b .t$; $y = b_2^2 3^3{}_{01}{}^t$; $.by = e^{(b0+b1.t)}$; and $(y = b_0 .e^{(b1.t)})$, obtaining the best results with the logarithmic model with $r^2 = 0.99$ and $S_{yx} \% = 8.6\%$.

Where: y=dependent variable; t=independent variable (time); b , b , b_{0123} =parameters , b; r^2=coefficient of determination; S_{yx} %=standard error of the mean in percentage.

The growth of *P. elliottii* in the region of this study was carried out previously by Tonini (2000) who, adjusting height curves as a function of age and using the Richards function with three coefficients, found a high coefficient of determination (0.99) and a low standard error of estimate in percentage (up to 10.6%), and for Piratini, RS, the dominant height was between approximately 17.5 and 22.5 meters at 15 years of age (Figure 2).

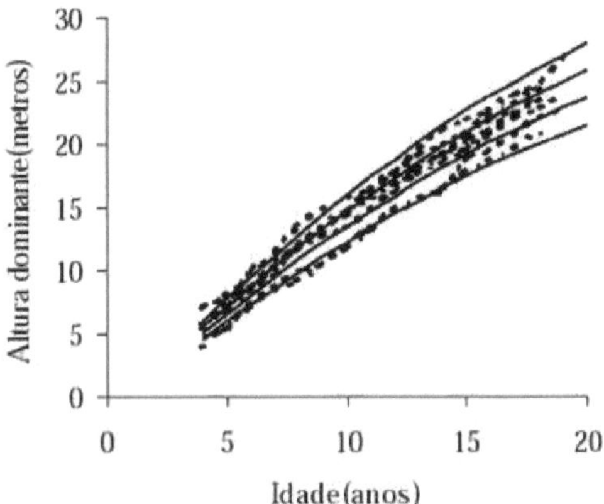

FIGURE 2 - Site index curves (4 filled lines) for *P. elliottii*, in Piratini. Source: Tonini (2000).

Tonini (2000) analyzed the covariance between various sites as a way of verifying the parallelism between the growth curves of *P. elliottii* using as effects: the calculated function (Richards equation), the sites, age and the interaction between age and site. The Richards equation is described as follows:

$$y = A\left\{\left[1 - e^{(-k.t)}\right]^{\,r}\right\}$$

Where: *A* = upper asymptote; *k* = growth rate, ranging from 0 to 1; *r* = inflection point of the curve, between the minimum value of y and the asymptote; y = dependent variable (dendrometric variables); t = independent variable (time).

The Forest Practices Branch of the Ministry of Forests of the Province of British Columbia (2001) lists the effects of pruning from different aspects as follows:

a) Physiological response of the trees: The reduction in the living canopy implies a reduction in the capacity to intercept light and, consequently, in photosynthetic capacity. Trees tend to maintain height growth, to the detriment of radial growth, due to the reduction in photosynthesis. The pruned trees continue to grow and the size of their crown is rebuilt, with height and diameter growth returning to normal. After pruning, each year the radial increment adds a layer of clean wood around the pollarded core. The radial increment is related to the size of the tree. One of the reasons why large old trees are not pruned is because they have passed the period of maximum annual increment in diameter and the layer of clean wood that can be expected is not sufficient to justify the investment in the operation. Many conifers have slow natural shedding and, as a result, produce a small amount of clean wood due to knots. Pruning alters the allocation of radial growth along the stem, reducing the taper of the pruned log over time.

b) Site response: In order to achieve a high rate of diameter growth in pruned trees and minimize the costs of pruning, the activity is generally carried out in wide spacings. Pruning causes a temporary reduction in the growth rate which is followed by a recovery to a quantity close to the stock without pruning. A new generation of trees can be established by natural regeneration (or released) in wide spacings of pruned trees;

c) Size of pruned trees: There is a relationship between the desire to minimize the size of the core, prune an adequate length of stem to harvest a clean log and retain enough live canopy to ensure rapid growth. Theoretically, dominant trees are more likely to accumulate clean wood than smaller trees, which also grow less.

d) Site quality: On the best sites, recovery to a "normal" level of growth is faster. This means that it is more economically viable to prune on more fertile sites than on poorer ones.

e) Fertilization: Fertilization can be used to improve the site and increase the speed of crown recomposition, recovering growth rates more quickly by adding greater quantities of clean wood.

f) Region: The environmental conditions of a given region have an impact on species, site quality and other factors such as forest health, which can be a limiting factor for pruning in regions where the growth of the species in question is slow.

g) Forest health: It is important to consider the interaction between forest health and silvicultural treatments. As pruned stands generally have a low stock of trees per unit area, additional mortality due to health factors has a major impact on production. Heavy pruning stresses the trees, which may become less able to respond to pest and disease attacks. Sudden exposure to direct sunlight can cause damage to the cambium. Cutting can allow insects to enter. Rot can penetrate the wounds of pruned branches. On some sites, pruning can cause an increase in root disease caused by *Armillaria* (fungus) in young stands. Pruning can also reduce losses to some pathogens such as rust in young stands, but it also has negative repercussions, as it is a trauma and affects the health of the trees (Emmingham and Fitzgerald, 1995).

h) Partial pruning: It is common to prune only some of the trees in the stand, usually in high-density plantations, which generally reduces the individual growth of the pruned trees; as a result, the unpruned trees can overtake the pruned trees.

In Tasmania, the price of *Pinus radiata* Don timber, pruned for export, reaches prices around 400% higher than unpruned timber, but pruning must be carried out at the right time, pruning before the season delays growth and pruning late means losing valuable timber that might not have knots (Beadle and Hall, 2001).

Kramer and Kozlowski (1972) state that pruning reduces growth in trunk diameter below the crown the most and that a reduction in crown height of up to a third of the tree's total height generally does not significantly affect tree growth; Schweingruber (1996) lists research that has shown that pruning in conifers can be carried out by removing lower branches up to 50% of the total crown length without reducing growth. Pruning *P. elliottii* to 40% of the total height of the trees implies a small reduction in production (Schneider, 2002). Mattos (sd) recommends that pruning be started when the plants are 6 m high and carried out up to around 30-40% of the total height of the plants. A general rule, accepted for all species and sizes of trees, is that at least 50 percent of the total height of the trees should be left in the living crown so as not to hinder growth (Emmingham and Fitzgerald, 1995). It is now common knowledge in forestry circles that pruning *P. elliottii* to 40% of the total height of the trees has no significant influence on production.

Pruning live branches, or green pruning, is always traumatic and pruning without scientific criteria can depreciate the value of trees and reduce their growth. However, applying knowledge of how the trees actually respond to the damage caused by pruning can lead to greater financial results; it should be borne in mind that branches that are already dead when cut do not involve direct damage and that, even if they are alive, the smaller they are, the less the trauma resulting from pruning (Letson, 1994).

Gibson *et al.* (2001) studied three treatments against a control in a *Pinus taeda* Linnaeus stand, the first of which was thinned, the second pruned and the third thinned and pruned. They concluded that the control had lower production in terms of log volume and that the wood from the pruned treatments was of better quality for lamination, resulting in higher net veneer production.

Schneider *et al.* (1999), studying the effect of pruning on *P. elliottii* planted in a region with poor soils in the Pinheiro Machado unit, in Cachoeira do Sul, Rio Grande do Sul (the same soil unit as the experiment in this study), with 5 treatments: control with no pruning, pruning of dry branches and pruning up to 12 m at intensities of 40%, 50% and 60% of the total height of the trees; at 11 years of age, they obtained averages of 263.5 m³/ha, 245.1 m³/ha, 231.5 m³/ha, 225.5 m³/ha, and 211.6 m³/ha, respectively. The control and pruning treatments did not differ from each other and were classified in group "A" in Duncan's test. The 40%, 50% and 60% pruned height treatments also did not differ and were classified in group "C" of Duncan's test. Group "B" of Duncan's test was made up of the dry pruning and 40% pruning treatments.

From these studies it is possible to deduce that denser stands produce more wood, the trees have fewer and smaller branches and a more cylindrical trunk, which is desirable for higher yields in lamination and sawmills, but the branches, especially the dead ones, must be removed to provide better quality wood.

Thus, it can be seen that there are benefits and harms associated with pruning and that the number of trees to be pruned is an economic issue that should be studied on a case-by-case basis, taking into account the time and place, making it possible to research and establish how much to prune on a technical-scientific basis in order to obtain a greater volume free of knots and, consequently, with better characteristics for carpentry.

3 MATERIAL AND METHODS

3.1 Study site

The property where the experiment was set up is called Fazenda Guanabara, located in the municipality of Piratini, state of Rio Grande do Sul, Brazil, 350 km from Porto Alegre and 380 km from Santa Maria, with approximate geographical coordinates of 53° 00' Greenwich west longitude and 31° 34' south latitude.

3.2 Regional characteristics

The region where the experiment is located is part of the geomorphological province of the South Rio Grande shield, with igneous rocks from the Precambrian period that have been badly eroded. It is part of the region of mixed and sub-shrub fields with the occurrence of gallery forests left over from the Serra do Sudeste, the latter being bordered to the north by the Central Depression, to the east by the Coastal Plain and to the south by the clean fields of the Serra do Sudeste (SCP, 2004). The soils are part of the Pinheiro Machado mapping unit, consisting predominantly of typical dystrophic neolithic soils (Streck *et al.*, 2002), derived from granite, well-drained, dark in color and medium in texture, generally acidic, well supplied with organic matter, with topographic gradients of eluvial soils at the top of the slopes, colluvial soils at mid-slope and alluvial soils in the lowlands (Tonini, 2000 and Higa *et al.*, 2003).

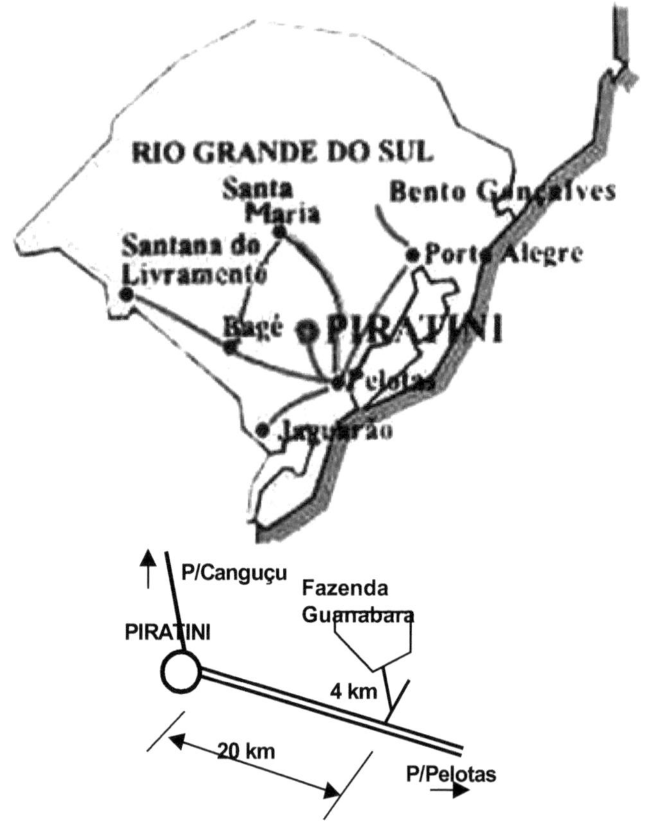

FIGURE 3 - Sketch of access to the property where the experiment is located (Fazenda Guanabara).

The region is part of the Crystalline Massif morphological complex, with undulating relief, dissected in the form of hills; it is located at an altitude of between 200 and 600 meters above sea level; the average annual temperature is around 16°C, the average of the coldest month around 12°C and the hottest month around 22°C (SCP, 2004), with an average of 23 frosts per year (Tonini,

2000; Higa *et al.,* 2003) and average annual rainfall of between 1500 and 1600 mm (SCP, 2004), which classifies the local climate as Cfb, according to the Koeppen system (Carvalho, 1994).

The municipality is bathed by the Camaquã and Piratini rivers, forming part of the Litorânea Hydrographic Basin. The Piratini River borders the property where the experiment is being carried out to the north.

3.3 Characteristics of the forest stand and the experiment

The area where the experiment was set up consists of a stand of *P. elliottii*, planted in 1985 with a spacing of 3 m x 2 m (6 m² of area per tree), made up of 6 plots totaling 251.69 ha. In 2000, a systematic thinning of the 6th planting line was carried out at a distance of 3 meters, complemented by thinning from below in the intermediate lines, with a total intensity of approximately 40% of the number of trees in the stand.

The experiment was set up in June 1991 and was located in plot 2. The experimental design was randomized blocks with 4 treatments, corresponding to a control and three pruning intensities, as shown in Table 2, with 4 replications represented by the four blocks.

TABLE 2 - Treatments applied from 1991 onwards in a *P. elliottii* stand in Piratini, RS

Code	Treatment	Description
0	Witness	without pruning
40	Treatment 1	pruning up to 40% of the total height of the trees
60	Treatment 2	pruning up to 60% of the total height of the trees
80	Treatment 3	pruning up to 80% of the total height of the trees

The plots (Figure 4) have dimensions of 20 m x 21 m (420 m^2), installed inside the stand, with a line of border trees. The total useful area of the experiment is 6.720 m^2 (4 plots x 4 blocks x 420 m^2) with a total of 1120 trees.

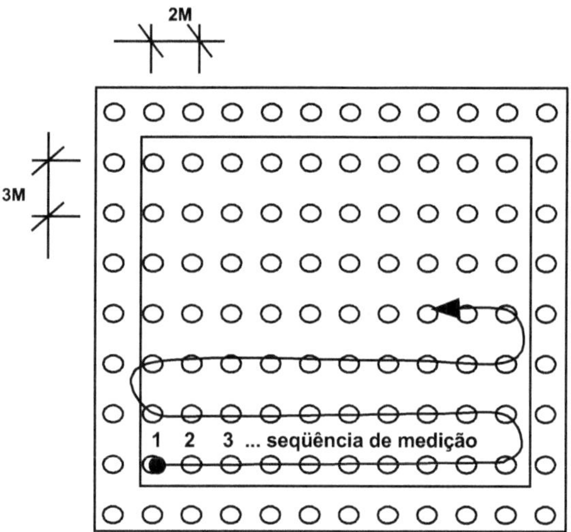

FIGURE 4 - Scheme for distributing the trees in the plots and how they were measured.

The height of the pruning was corrected year by year, up to a maximum height of 6 meters, starting in 1991 at the age of six (1st measurement) and continuing until 1999 (9th measurement), when all the trees in the experiment were pruned to a height of 6 meters (Figure 12).

The distribution of the blocks and plots in the experiment is shown in Figure 5, with the distribution of treatments per block; the summary history of the experiment is shown in Table 3.

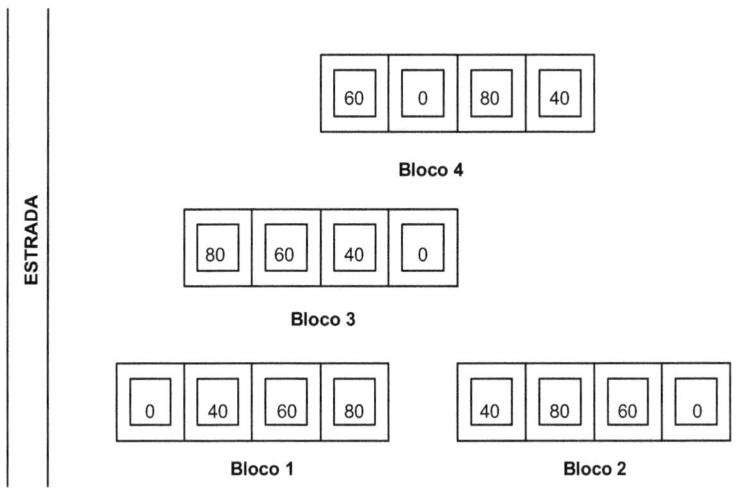

FIGURE 5 - Distribution of treatments by block.

3.4 Measured and estimated variables

From the 1st to the 6th measurement, the following were measured on all the trees in the useful area of each plot: the trunk diameter (d), taken at 1.3 m from the ground with a precision of 0.5 centimeters, using a suta, and the height (h) of all the trees using a hypsometer, with a precision of tenths of a meter. From the 7th to the 10th measurement, the diameter of all the trees was measured to the nearest 0.5 centimeter using a suta, and the height (h) of the

27

first 10 trees was measured using a hypsometer to the nearest tenth of a meter.

The experiment was thinned at 15 years of age, shortly after the 10th measurement and again at 18 years of age. The thinning did not follow technical criteria; the weight of the intervention and the distribution of the trees felled were not homogeneous, which is why the experiment was based on the age of 15 for analysis and inferences. The trees felled at 18 years of age for cubing and trunk analysis purposes were selected taking into account, in addition to the desired size, their spatial distribution among the remaining trees in order to avoid selecting trees with a larger living space due to thinning.

TABLE 3 - Experiment history and measurements

History	Date	Age (years)	Observations
Planting	1985	0	Planting the forest.
1st Measurement	11/06/1991	6	Measurement of diameter (d) and height (h) of all the trees in the plots.
Installation	06/1991	6	1st pruning - Setting up the experiment.
2nd Measurement	10/07/1992	7	2nd pruning and measurement of d and h of all trees.
3rd Measurement	09/09/1993	8	3rd pruning and measurement of d and h of all trees.
4th Measurement	08/1994	9	4th pruning and measurement of d and h of all trees.
5th Measurement	09/09/1995	10	5th pruning and measurement of d and h of all trees.
6th Measurement	24/10/1996	11	6th pruning and measurement of d and h of all trees.
7th Measurement	08/1997	12	7th pruning and measurement of d of all trees and h of the first 10 in each plot.
8th Measurement	29/10/1998	13	8th pruning and measurement of d of all trees and h of the first 10 in each plot.
9th Measurement	30/09/1999	14	9th pruning and measurement of d of all trees and h of the first 10 in each plot.
10th Measurement	11/2000	15	10th measurement of d of all trees and h of the first 10 in each plot.
1st thinning	2000	15	Systematic from the 6th line and selective in the middle.
2nd thinning	2003	18	Selective; applied from April to May in blocks 1 and 2.
11th Measurement	08/2003	18	11th measurement of d, h.
Cubage	11/2003	18	Rigorous cubing and trunk analysis, felling of 3 trees per plot.

3.4.1 Useful area of the plots (a)$_i$

The area of the plot was calculated by multiplying its width by the average length of the planting rows, with the boundaries determined by the average distance between the last row of trees in the useful area and the border (Figure 6).

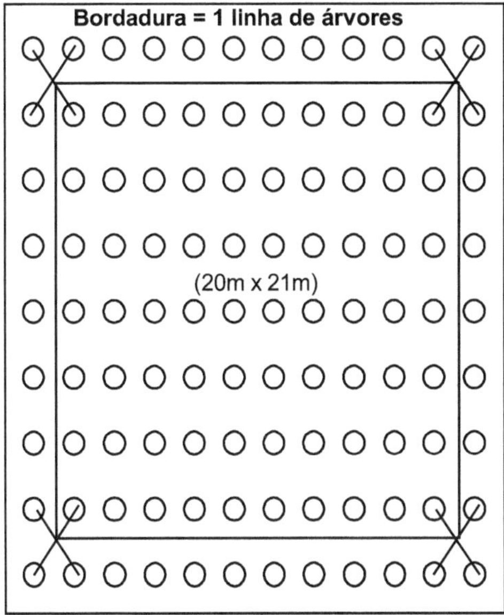

FIGURE 6 - Approximate useful area of the plots.

The useful area of the plots ($_i$) is:

$$a_i \cong \text{width x length} \cong 20 \text{ m x } 21 \text{ m} \cong 420 \text{ m}^2.$$

The number of trees per hectare (N_{ha}) in each plot was estimated using the equation:

$$N_{ha} = 10000.N_i / a_i .$$

Where: N_i = number of trees in plot i at the time of measurement; a_i = area of plot i in m².

3.4.2 Mortality (dead%)

To analyze the effect of pruning on mortality, only dead plants after the start of pruning at 6 years of age were considered.

The analysis of variance for mortality was carried out with the values transformed as recommended by Gomes (1982) for small percentage values, as follows:

$$Y= \sqrt{Mortas\% + 0{,}5}$$

Where: Y=dependent variable analyzed; Dead%=number of dead trees per plot, in percentage, after 6 years of age.

3.4.3 Diameter (d)

The diameter at breast height (d), taken at a height of 1.3 m, was measured with a stylus, directly on all the trees in the useful area of each plot, from the 1st to the 10th measurement, taking two diameters orthogonally and noting the average of the two with a precision of 0.5 cm.

3.4.4 Height (h)

Height (h) was measured on all the trees in each plot from the 1st to the 6th measurement, with an accuracy of 0.10m.

From the 7th to the 10th measurement, the heights of the first ten trees in each plot were taken with an accuracy of 0.10 m. Only these trees were used in the analysis of variance of the average height per plot.

A large number of models have been developed to estimate height as a function of diameter and many researchers have tested them, including Machado *et al.* (2000), who chose 4 different models to adjust the hypsometric relationship equations for *Araucaria angustifolia* (Bert.) O. Ktze, the parabolic model ($h = b_0 + b_1 d + b_2 d^2$), Stofells (*Ln h = b_0 + b_1 ln d*), Curtis *Ln h = b_0 + b_1 (1/d)* and Prodan $h = d^2 / (b_0 + b_1 d + b_2 d^2)$, finding the best results with the Stofells model; where: *h the* height (m); *ln* the Neperian logarithm of the variable considered; *d* the diameter (cm); b_0, b_1 *and b_2* the parameters of the equations.

Using the independent variables from the equations in the previous paragraph, we obtain the 9 models listed in Table 4, cited by Finger (1992), which were tested to describe the hypsometric relationship.

TABLE 4 - Equation models tested to estimate the height (h) of *P. elliottii* trees in Piratini, RS

No. eq.	Equation	Author
1	$h = b + b_{01} .d + b_2 .d^2$	-*
2	$ln\ h = b + b_{01} .ln\ d$	Stofells*
3	$ln\ h = b + b_{01} (1/d)$	Curtis*
4	$h = b + b_{01} (1/d)$	-
5	$h = b + b_{01} .d + b_2 (1/d)$	-
6	$h = b + b_{01} (1/d) + b_2 .d^2$	-
7	$h = b + b_{01} .d + b_2 (1/d) + b_3 .d^2$	-
8	$h = d^2 /(b + b_{01} .d + b_2 .d)^2$	Prodan* (non-linear)
9	$h = b + b_{01} .ln\ d + b_2 .ln\ d^2$	-

Where: h=height; ln=Neperian logarithm; d=diameter. b, b, b_{012} and b_3 =equation parameters. Source: (*) Machado *et al.* (2000).

3.4.5 Dominant height (h)$_{100}$

The dominant height in this study was defined as the arithmetic mean height of the 100 thickest trees per hectare, or Assmann's dominant height (Finger, 1992).

From 11 to 15 years of age, the trees that were not measured had their height calculated by the hypsometric relationship using equation no. 8, or Prodan's model (Table 4), as only the first 10 trees in each plot were measured at those ages.

3.4.6 Basal area

The individual basal area (g) of each tree, in square meters, was calculated using the equation:

$$g = \pi \cdot d^2 / 4$$

Where: g = individual basal area (m²); π =3.141593; d=diameter (m).

The basal area per hectare per plot (G_i) was determined by multiplying the number of trees per hectare by the average individual basal area ($\overline{g_i}$) of the trees in plot i. The annual current increment (ICA) in basal area per hectare (m²/ha/year) was determined by the difference between the basal area per hectare of plot i, at the age considered, and that presented in the previous year.

3.4.7 Cubage

In the 11th measurement (2003), a sample of felled trees was taken, cubed and subjected to trunk analysis according to the method described by Husch *et al.* (1982), with the aim of estimating volumes and volumetric increments. Three trees were sampled per plot: a) the first tree - with a diameter at 1.3 m (d) approximately corresponding to the tree with the average basal area of the plot ($\overline{g_i}$); b) the second tree - with d approximately corresponding to $\overline{g_i}$ +1s; c) the third tree - with d approximately corresponding to $\overline{g_i}$ -1s;

where s is the standard deviation of the basal area of the trees in the plot.

The individual volume (v) of the felled trees in cubic meters was determined using the Smalian method, according to Finger (1992):

$$v_t = v_0 + \sum_{i=1}^{n} v_i + v_c$$

Where: v_0 = volume of the stump (m^3); v_i = volume of the intermediate sections (m^3) calculated by multiplying the length of the log by half the sum of the basal areas of the two ends of each section; v_c = volume of the upper end of the trunk (m^3) calculated using the cone formula with a length equal to that of the section considered.

The individual volumes (m^3) of the standing trees were estimated using the Stoate equation (Table 5, p. 25).

3.4.8 Trunk analysis

The 48 trees sampled for volume calculation were sectioned into logs of the standard length of 2.8 m used commercially on site. A disk was extracted from each tree at a height of 1.3 m from the ground, a disk from the base of each section and a disk from the top of the last section, all 4 to 8 cm thick, for trunk analysis. The disks were dried in an oven and then their circumferences were measured with and without bark. Four radii were taken from each disk, the first at an angle of 45° from the largest radius of the slice and the others at 90° from each other. The growth rings were identified using a magnifying glass and their thicknesses were measured using a magnifying glass fitted with a micrometric measuring table coupled to a computer.

Growth rings were identified in accordance with Schweingruber (1996) and Larson *et al.* (2001) who describe the problems in identifying annual growth rings in *Pinus*, citing that there can be very faint rings close to the pith that are difficult to identify visually and false rings caused by stress, listing the factors that influence their formation and some techniques used to identify true rings.

In the trunk analysis, the annual rings were not always visible, and the innermost rings were identified mainly by their resistance to compression, since in most cases there was no visual differentiation by the color or texture of the wood of the first three or four rings, even at maximum magnification. The first three slices from the base onwards had a greater number of rings that were difficult to identify. In these cases, a steel blade about 0.5 mm thick was used to compress the wood along the radius of the slice and the rings formed were considered to be resistant to compression.

Once the thickness was measured, the total volume with and without bark at the last age, the length of the tips at each age, the total volume without bark by age and the volumes of the first two logs 2.8 m long from 0.2 m high (height of the felling cut) with and without knots were calculated.

The heights of the trees felled at each age were determined by summing the lengths of the sections of the corresponding layer, adding the length of the upper tip of the stem and the height of the stump (0.2 m). The inner tips were estimated by interpolation, as shown in Figure 7.

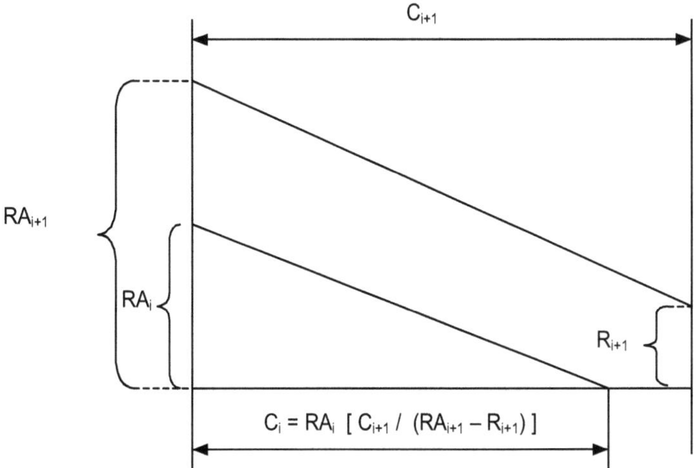

Where: i = considered age (years); C = length (m); RA = radius in the previous section (m); R = radius in the current section (m).

FIGURE 7 - Calculation of tree tip length by interpolation, at each age, in the trunk analysis.

3.4.9 Volume

The individual volume of clean wood on the corrugated core of the trees felled in the pruned treatments was calculated by reducing the volume of the corrugated core from the total volume of each section individually (first and second log). The diameter of the core of the first log (d_{e1}) was taken to be the diameter at the base of the log (0.2 m) at the age at which pruning reached a height of 3.0 m. For the second log, the diameter of the cored core (d_{e2}) was taken to be that measured at a height of 3.0 m at the age when the pruning height reached 5.8 m. The length of the first and second logs was 2.8 m. The volume of the first log was calculated as the sum of two subsections: the first subsection starts at the base of the log at a height of 0.2 m from the tree and goes up to a height of

1.3 m, while the second subsection goes from a height of 1.3 m up to 3.0 m.

In this study, the equation models listed in Table 5 were tested to estimate individual volume (m³) as a function of tree diameter (cm) and height (m).

TABLE 5 - Equation models tested to estimate the volume of *P. elliottii* trees in Piratini, RS

No. eq.	Equation	Author
1	$v = b_0 + b_1 . d$	-
2	$v = b_0 + b_1 . d^2$	Kopezky-Gehrhardt
3	$v = b_0 + b_1 . d\ h^2$	Spurr
4	$v = b_0 + b_1 . d^2 + b_2 . d^2 . h + b_3 . h^3$	Stoate
5	$\ln v = b_0 + b_1 . \ln d$	Hush
6	$\ln v = b_0 + b_1 . \ln d^2 . h$	Spurr - logarithmic
7	$\ln v = b_0 + b_1 . \ln d + b_2 . \ln h$	Schumacher-Hall
8	$\ln v = b_0 + b_1 . \ln d + b_2 . (\ln d)^2 + b_3 . \ln h + b_4 . (\ln h)^2$	IPF Baden-Wurt, Germany

Where: v=volume; h=height; log=logarithm in base 10; ln=Neperian logarithm; d=diameter. b_0, b_1, b_2, b_3, b_4 = parameters of the equations.
Source: Loetsch *et al.* (1973).

3.4.10 Growth

To describe the growth of trees from 6 to 18 years old, 12 equation models were tested (Table 6, p. 26).

TABLE 6 - Equation models for describing the growth of *P. elliottii* trees in Piratini, RS

No. eq.	Equation	Author / type
1	$y = b_0 + b_1 . t$	-*
2	$y = b_0 + b_1 . \ln t$	-*
3	$\ln y = \ln(b_0) + b_1 . \ln t$	-*
4	$y = b_0 + b_1 / t$	-*
5	$y = b + b_{0\,1} . t + b_2 . t^2$	-*
6	$y = b + b_{0\,1} . t + b_2 . t + b^2{}_3 . t^3$	-*
7	$y = b_0 . t^{\,b1}$	-*
8	$y = e^{(b0 - b1.t)}$	-*
9	$y = b_0 . e^{(b1.t)}$	-*
10	$y = A\left\{\left[1 - e^{(-k.t)}\right]^r\right\}$	Chapman-Richards**
11	$y = b_0 / (b_3 + e^{(b1 - b2.X)})$	Logistics *
12	$y = b_0 . e^{e^{(b_1 - b_2.X)}}$	Gompertz*

Where: y = dependent variable (diameter and height); t = time in years; $\ln$ = Neperian logarithm; e = base of the Neperian logarithm; b , b , $b_{012,}$ b_3 = parameters of the equations; A, k, r = parameters of the Chapman-Richards equation. Sources : (*)Sit (1994); (**) Prodan (1997).

3.4.11 Tree shape

The analysis of shape variation between treatments was carried out in accordance with Finger (1992) and Scolforo *et al.* (1998), using the artificial shape factor (f), calculated using the equation:

$$f = vr / vc$$

Where: $v =_r$ Strict stem volume (m³) calculated using the Smalian method, v_c = Cylinder volume (m³) with length (m) equal to the total height of the tree and diameter d, where d=diameter (m) taken at 1.3 m from the ground.

The shape was also studied using the equation model:

$$gx = p.x^r$$

Where: r >zero, representing the shape of the tree, being more cylindrical the closer r *is to* zero; x = distance (m) from the upper end of the stem, g_x is the basal area (m²) at the distance considered and p is the parameter describing the size of the rotation body.

3.5 Spacing

In order to assess the living space and as an indication for thinning, the relative spacing index (S%) was calculated which, according to Schneider (2002), was estimated by:

$$S\% = 100.EMi/h0$$

Where: h_0 = dominant height (m); MS_i = average linear spacing (m) per tree in plot i, calculated by $MS_i = \sqrt{a_i/N_i}$, where "a_i" is the area (m²) and "N_i" is the number of trees in plot i.

The dominant height used in this work was the average of the 100 thickest trees per hectare (h_{100}).

Fishwick (1975) *apud* Finger and Schneider (1999), in studies on thinning *P. elliottii*, determined that basal area production is maximum when the S% reaches 21% and competition is severe when the S% drops to 16%, recommending that a stand at first thinning age should increase the S% to 21%. According to the results of the Continuous Forest Inventory of Rio Grande do Sul (UFSM/SEMA-RS, 2003), the *Pinus* forests of the Serra do Sudeste in Rio Grande do Sul showed a population density index (= relative spacing index), calculated by the ratio between average spacing and dominant height, with an average of 16%, ranging from 11.2% to 27.4%, and in 62% of the plots the value was less than or equal to 16.0%, characterizing stands where the trees are considered to be in competition, and the average basal area was 37.4 m² /ha, ranging from 15.00 m² /ha to 58.9 m /ha.²

3.6 Analysis of variance and mean comparisons

Due to the thinning that began at 15 years of age, the analysis of the experiment was limited to this age. Therefore, the analysis of differences between treatments and the selection of equation models were based on the data collected at 15 years of age.

The experimental design was randomized blocks (4 blocks) with 4 treatments and the following model, according to Storck and Lopes (1998):

$$Yij = m + ti + bj + eij$$

Where: Y_{ij} = response variable (measured or estimated value); m = constant mean; t_i = treatments; b_j = blocks; and$_{ij}$ = error.

Analysis of variance was carried out on measurements from 6 to 15 years of age using diameter with bark at breast height (d), height (h), mortality, basal area per hectare (G) and the relative spacing index (S%) as response variables.

The analysis of variance, using the total individual bark-free volume (v_{sc}) and the artificial form factor (f) as response variables, was carried out only on the data collected at 15 years of age.

3.7 Pearson's correlation

The correlation study between the variables studied, when necessary, was carried out using Pearson's correlation coefficient given by (Wonnacott and Wonnacott, 1980):

$$r = \frac{\sum xy}{\sqrt{\sum x^2 \; \sum y^2}}$$

Where: r = Pearson's correlation coefficient; x, y = variables under study.

The calculations were carried out using the PROC CORR procedure of the SAS System (SAS Institute, 2001).

3.8 Data normality test

One of the presuppositions for applying the F and mean tests in the analysis of variance is that the data has a normal distribution. The Shapiro-Wilk test was therefore applied to the diameter distribution of the trees in the useful area of the plots at 15 years of age.

The Shapiro-Wilk W statistic (IU, 2004) is the ratio between the best variance estimator and the sum of the corrected squares of the variance estimator of the data collected. The value is positive and less than 1, being closer to normality the closer it is to 1. The W statistic requires values between 7 and 2000 units, being the standard for small samples ($\leqslant$ 2000) through the SAS System's UNIVARIATE procedure, which uses the Kolmogorov-Smirnov test for sample sizes greater than 2000 units. A significant value for W indicates a lack of normality for the variable analyzed (Anjos, 2003). The W value is calculated as follows (IU, 2004):

$$W = \frac{\sum \left(a_i x_{(i)}\right)^2}{\sum \left(x_i - \bar{x}\right)^2}$$

Where $a_i = (a, a_{12}, ..., a_n) = m'.V^{-1} [m'.V^{-1}.V^{-1}.m]^{-1/2}$; $m' = (m, m_{12}, ..., m_n)$ is the vector of expected values of the normal-order statistic; V is the n-by-n covariance matrix; $x' = (x_1, x_2, ..., x_n)$ is a random sample and $x_{(1)} < x_{(2)} < ... < x_{(n)}$.

3.9 Criteria for selecting regression equations

Couto and Vettorazzo (1999) selected equations for the volume and weight of *P. taeda* using the following procedure: (a) examination of the analysis of variance table: sum of squares of the residuals (SQR), mean square of the residuals (QMR), F-test for the complete model and sequential F-test; (b) analysis of the precision measures: coefficient of determination (R^2), standard error of the estimate or residual standard error (S_{yx}) and standard error of the estimate expressed as a percentage of the arithmetic mean of the dependent variable ($S_{yx}\%$), interpreted in a similar way to the coefficient of variation (CV%); (c) graphical distribution of the residual values and (d) examination of the estimates of the equation's parameters using the t-test. Abreu *et al.* (2002) used graphical analysis of the residuals, the standard error of the estimate (S_{yx}) and the coefficient of determination (R^2) to select the best volume equation by diametric class for *Eucalyptus grandis*. Machado *et al.* (2000) used the adjusted coefficient of determination (R^2 aj.), the standard error of the estimate in percentage ($S_{yx}\%$) and the graphical distribution of the residuals to select regression equations for total volume per unit area.

In this study, the equations were selected based on the best graphical distribution of the residues, the lowest coefficient of variation ($CV\%$) and the highest coefficient of determination (R^2). In the case of equations with similar results, the one with the fewest parameters was chosen. In a second stage, tests were carried out to determine the validity of the model previously selected as the best.

The graphical analysis of the residuals was carried out by observing their distribution against the values of the estimates according to Bussab (1986). The value of the residuals is calculated using the formula:

$$e_i = y_i - \hat{y}_i$$

Where: e_i = residual of observation i; y_i = observed value; $\hat{y}_i$ = estimated value.

Poor distribution of the residuals was considered to be: the formation of distribution patterns (transgressions), the concentration of the residuals above or below the axis of the estimates or their concentration in up to two-fifths of the amplitude of the same axis. Reasonable: the equitable distribution of the residuals above and below the axis of the estimates, in at least half the amplitude of the same axis and without the formation of distribution patterns. Good: the distribution of the residuals over 3 to 4 fifths of the amplitude of the axis of the estimates, equally above and below the same axis and without the formation of patterns. Only the distribution of residuals over more than four-fifths of the amplitude of the estimation axis, equally above and below

44

the axis and without the formation of patterns was considered optimal.

In the case of the hypsometric relationship and volume modeling, an analysis of covariance was carried out with regard to the treatments to check whether separate regression equations should be used for each treatment, as described in section "3.11 Analysis of covariance".

The coefficient of determination is given by the quotient between the sum of squares of the regression and the sum of squares of the total:

$$R^2 = SQreg / SQtotal$$

Where: R^2=*coefficient* of determination; SQ_{reg} =sum of squares of the regression; SQ_{total} =sum of total squares.

When necessary, in cases where the number of parameters differed between the best models, the R^2 was adjusted according to Schneider (1998) to enable the equation to be chosen:

$$R^2_{aj.} = R^2 - \left(\frac{k-1}{N-k} \right) \cdot \left(1 - R^2 \right)$$

Where: $R^2_{aj.}$ = adjusted coefficient of determination; *k=number of parameters* in the equation; *N=number of* observations.

The coefficient of variation (CV%) is the percentage represented by the standard error of the mean (S_{yx}) in relation to the mean of the observations of the dependent variable ($\bar{y}$), the standard error of the mean being given by the square root of the mean square of the residuals (QM)$_{res}$:

$$Syx = \sqrt{QMres}$$

$$CV\% = 100 \cdot S_{yx} / \bar{y}$$

To compare the logarithmic equations with normal linear functions, the QM_{res} was calculated by extracting the antilogarithm of the original dependent variable (y) and the estimated one ($\bar{y}$), according to Sit (1994).

3.10 Validation of regression equations

The tests used to validate the regression equations were carried out using the SAS System statistical package according to the procedures described by SAS Institute (2004).

The selected equations were validated by determining:

- Homoscedasticity of variance using White's χ^2 test;

- The independence of the residuals using the Durbin-Watson test;

- The normality of the distribution of residuals using the Kolmogorov-Smirnov or Shapiro-Wilk tests.

3.10.1 Homoscedasticity of variance

One of the main assumptions for the usual least squares regression is homogeneity of variance (homoscedasticity). If the model is well-fitted, there should be no pattern to the residuals plotted against the fitted values. If the variance of the residuals is not constant, then the residual variance is said to be "heteroscedastic". There are graphical and non-graphical methods for detecting heteroscedasticity. A commonly used graphical method is to plot the residuals against fitted values, as described in section "3.9 Criteria for selecting regression equations". The SAS System calculates the residuals and fitted values using the GLM, REG and NLIN procedures, which can be displayed on a graph. When the residuals are distributed without any pattern, there is no heteroscedasticity.

A mathematical method for determining whether there is homogeneity of variance in the residuals, which can be carried out using the SAS System, is White's test (SAS Institute, 2004). White's test is computed by finding nR^2 from a regression of e_i^2 on all distinct variables in $X \times X$, where X is the vector of dependent variables including a constant. This statistic is distributed asymptotically as Chi-squared (χ^2) with k-1 degrees of freedom, where k is the number of regressors.

The method tests the null hypothesis that the residual variance is homogeneous. Then, if the "p" value is too small, the hypothesis is rejected and the alternative hypothesis that the variance is not homogeneous is accepted. This is done by using the "SPEC" option in the model statement, as shown in the following example:

```
PROC REG;
   MODEL Y = X / SPEC;
```

The test can also be carried out using the WHITE option in the FIT statement of the MODEL procedure in the SAS System, as in the following example:

```
PROC MODEL;
```

```
PARMS A B C;
Y = A + B * X1 + C * X2;
FIT Y / WHITE;
```

3.10.2 Independence of residues

The value of the Durbin-Watson "d" statistic (SAS Institute, 2004) is obtained using the CLM option of the MODEL statement of the GLM procedure of the SAS System, or the DWPROB option of the FIT statement of the MODEL procedure, or the DW option of the MODEL statement of the REG procedure, as shown in the example below:

```
PROC REG;
   MODEL Y=X1 X2 / DW;
```

The "d" statistic is expected to be approximately equal to 2 if the residuals are independent. Otherwise, if the residuals are positively correlated, it will tend to be close to 0 (zero), or close to 4 if the residuals are negatively correlated (Nemec, 1996).

The value of d is given by:

$$d = \frac{\sum_{i=2}^{n}\left(E_i - E_{i-1}\right)^2}{\sum_{i=1}^{n}E_i^{\;2}}$$

Where: d = Durbin-Watson "d" statistic; E_i = stochastic error $= \hat{Y}_i - Y_i$; n = number of observations; $\hat{Y}_i =$ estimated value; Y_i = observed value.

3.10.3 Normality of the distribution of residuals

The principle of this test is based on comparing the cumulative frequency curve of the data with the theoretical distribution function hypothesized. When the two curves overlap, the test statistic is calculated as the maximum difference between them. The magnitude of the difference is established according to the probability distribution of this statistic, which is tabulated. If the experimental data deviates significantly from what is expected from the hypothesized distribution, then the curves obtained should be equally distant and, by analogous reasoning, if the fit to the hypothesized model is admissible, then the curves develop closely.

The Kolmogorov-Smirnov statistic (D) (SAS Institute, 2004) is an Empirical Distribution Function (EDF) statistic. The Empirical Distribution Function (EDF) is defined for a set of n independent observations X_1 ,... , X_n with a common distribution function F(x). Under the null hypothesis, F(x) is the normal distribution. The observations are ordered from smallest to largest as X(1),... , X(n).

The empirical distribution function Fn(x) is defined as:

$F_n (x) = 0, x < X_{(1)}$

$F_n (x) = i/n, X_{(i)} \leq x < X_{(i+1)} , i = 1,2,...,n-1$

$F_n (x) = 1, x \leq x_{(n)}$

Note that Fn(x) is a sequential function that advances in [1/n] with each observation. This function calculates the distribution function F(x). At any value x, Fn(x) is the proportion of observations less than or equal to x, while F(x) is the probability of an observation being less than or equal to x. EDF statistics measure the discrepancy between Fn(x) and F(x). The computational formulas for EDF statistics make use of the transformation of the probability integral U=F(X). If F(X) is the distribution function of X, the random variable U is uniformly distributed between 0 and 1.

Given n observations of X(1),... , X(n), the values U(i)=F(X(i)) are computed as shown below. The Kolmogorov-Smirnov statistic (D) is based on the largest vertical difference between F(x) and Fn(x), and is defined as:

$D = \sup_x |F_n(x) - F(x)|$

The Kolmogorov-Smirnov statistic is computed as the maximum of D+ and D-, where D+ is the greatest vertical distance between the EDF and the distribution function when the EDF is greater than the distribution function and D- is the greatest vertical distance when the EDF is less than the distribution function.

$D^+ = \max_i \left((i/n) - U \right)_{(i)}$

$D^- = \max_i \left(U_{(i)} - (i-1)/n \right)$

$D = \max \left(D^+ , D^- \right)$

SAS System's CAPABILITY procedure uses the modified Kolmogorov D statistic to test the data against the normal distribution with mean and variance equal to the sample mean and variance. In the MODEL procedure, the statistic is only used for samples over 2000 individuals. In the case of small samples, the Shapiro-Wilk test described in section "3.8 Data normality test" is used instead of the Kolmogorov-Smirnov test. In the MODEL procedure, the normality test is obtained using the NORMAL option in the FIT statement, as in the example:

```
PROC MODEL;
  PARMS A B C;
  Y=A+B*X1+C*X2;
```

FIT Y / NORMAL;

3.11 Analysis of covariance

Analysis of covariance was carried out to verify the need to use independent functions per treatment to describe the hypsometric and volume relationship, according to the procedures described by Schneider (1998) and Freund *et al.* (1986).

According to Storck and Lopes (1998), when calculating covariance it is possible to check the assumption of homogeneity of the estimated linear regression coefficients ($\hat{\beta}$) by calculating a coefficient separately for each treatment and then testing their equality.

Schneider (1998) uses the "F" value of the interaction of the covariate with the estimated dependent variable ($\hat{y}$) for each treatment, calculated by the sum of squares of corrected products, to compare the differences between slopes and uses the "F" value calculated for the covariate through the sum of squares of corrected products of the model without interaction to compare the difference between levels when there is no difference between slopes. In the SAS System, the sums of squares of corrected products are obtained using the SS3 option in the MODEL statement of the GLM procedure (SAS Institute, 1985). Treatments were defined as the covariate when checking for the need for individual regression equations per treatment.

3.12 Modeling diameter and volume as a function of pruning intensity

The variables diameter (cm) and volume without bark (m³) were modeled as a function of pruning intensity, testing the partial F of the linear, quadratic and cubic effects of the treatments using the model:

$$y = b_0 + b_1 .X + b_2 .X^2 + b_3 .X^3$$

Where: y=dependent variable (*d, v*); X=independent *variable* (pruning intensity in percentage); b , b , b_{0123} , b= parameters.

4 RESULTS AND DISCUSSION

4.1 Results on standing trees

4.1.1 Diameter at breast height (d)

The only variable measured on all occasions, on all the trees in the useful area of the plots, was diameter (d), which was the main control variable in the experiment. Therefore, data normality analysis was carried out on this variable at the age of 15, when it was considered that there was the maximum effect of the treatments without the effect of other types of silvicultural interventions.

The normality analysis of the diameters at 15 years of age, using the Shapiro-Wilk test, showed non-significant values very close to 1 (Table 7); this indicates that the frequency distribution is close to normal (Figure 8).

TABLE 7 - Results of the Shapiro-Wilk test on the diameters of the trees in the useful area of the plots at 15 years of age, by treatment

Treatment ->	General	0%	40%	60%	80%
W : Normal	0,983572	0,985931	0,976962	0,979094	0,980734
Pr<W	0,0789 ns	0,7211 ns	0,1141 ns	0,1907 ns	0,2961 ns

Where: Pr<W: probability of significance of W. NS= not significant at the 5% probability level.

Once the normality of the data was confirmed, the applicability of the significance tests based on this type of distribution was guaranteed.

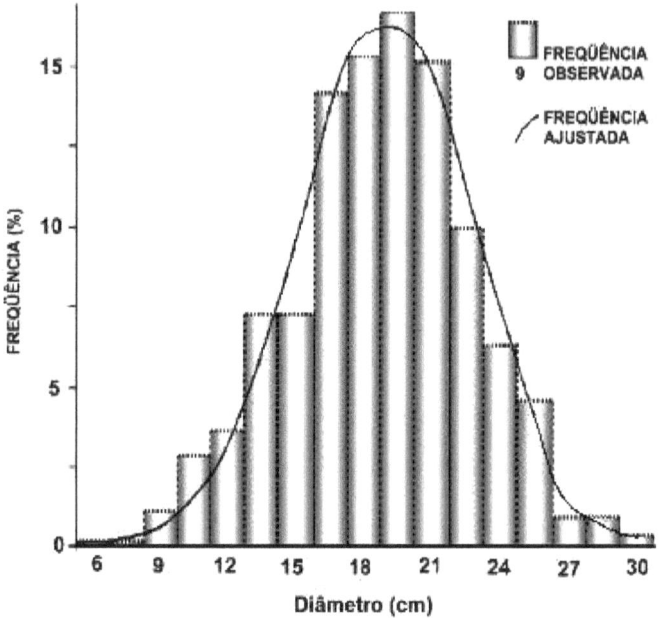

FIGURE 8 - Normality of the frequencies by diameter class, of the trees contained in the useful area of the plots, at 15 years of age.

Among the dendrometric variables measured, diameter was the one most affected by pruning. Before pruning, the control treatment had a significantly lower average diameter than the other treatments. This situation was reversed one year after the pruning treatments were applied, as can be seen from the data in Table 8, plotted in Figure 9. Pruning had a significant effect at the 1% probability level on diameter from 7 to 15 years of age, according to the F test. The treatment with 40% of the height pruned resulted in a 2.0% reduction in diameter compared to the control treatment with no pruning, with no difference between the diameter averages of the two by Tukey's test at 5% probability. When pruning was increased to 60% of the height, the reduction in diameter reached 7.5%, less than the control treatment according to the Tukey test at 5% probability level. At 80% of pruned height, the reduction is 13.5%, with the average diameter lower than all the other treatments according to the Tukey test at 5% probability level. These results are somewhat similar to those obtained by Schneider *et al*. (1999) in a pruning experiment with *P. elliottii, in which* they found a reduction in diameter of 4.9% in the 40% pruning treatment compared to the treatment without pruning and 10.4% in the 60% pruning treatment.

Table 8 shows that the mean square of the error increased with age, showing that variations within plots increased over time, resulting in a CV of 19.8% at 15 years of age, serving as an indication for a greater number of repetitions in future experiments.

TABLE 8 - Evolution of average diameter (cm) with age (years) by pruning intensity and statistical tests, for *P. elliottii*, in Piratini

Age (years)	Pruning treatments and Tukey's test							QM_{res}
	0 %	40 %		60 %		80 %		
	d (cm)	d (cm)	Dif%	d (cm)	Dif%	d (cm)	Dif%	
6	5,9b	6,0a	1,7	6,4a	8,5	6,4a	8,5	3,32
7	9,2a	9,1a	-1,1	8,7b	-5,4	8,2c	-10,9	3,55
8	12,0a	11,5b	-4,2	11,0c	-8,3	9,5d	-20,8	3,86
9	14,3a	13,7b	-4,2	12,7c	-11,2	10,8c	-24,5	4,51
10	16,1a	15,5a	-3,7	14,2b	-11,8	12,3c	-23,6	5,81
11	17,0a	16,5a	-2,9	15,2b	-10,6	13,5c	-20,6	7,13
12	17,7a	17,4a	-1,7	16,1b	-9,0	14,5c	-18,1	8,67
13	18,6a	18,2a	-2,2	17,0b	-8,6	15,7c	-15,6	10,80
14	19,2a	18,8a	-2,1	17,7b	-7,8	16,5c	-14,1	11,84
15	20,0a	19,6a	-2,0	18,5b	-7,5	17,3c	-13,5	13,94

Where: d = diameter (cm); a, b, c, d = Tukey's test groups at the 5% probability level, horizontally; Dif% = percentage difference in relation to the 0% pruning treatment; QM_{res} = mean square of the residual.

Due to the significance of the effect of the treatments on the diameter at 15 years of age, when the maximum effect of the treatments was obtained without the additional effect of the thinning carried out shortly after the measurement at that age, the diameter was modeled as a function of the effect of the treatments, and significance was found at the 1% probability level by the F test for the linear (F=67.61) and quadratic (F=9.95) effects. The result was the equation in Table 9, plotted in Figure 10, which shows that the greater the intensity of pruning, the greater the reduction in diameter growth.

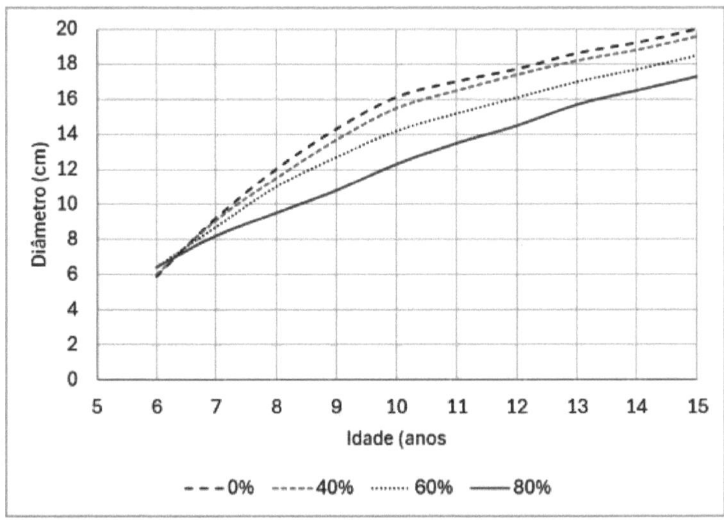

FIGURE 9 - Evolution of average diameter (cm) with age (years) by pruning treatment.

TABLE 9 - Equation for estimating diameter (cm) as a function of treatment at 15 years of age ($d=b+b_{01} \cdot Trat+b_2 \cdot Trat^2$)

Parameter	Value	Statistics	Value
b_0	20,025	R^2	0,90
b_1	0,00754	$\bar{d}$	16,5
b_2	-0,00052	CV%	20,3

Where: b_0 , b_1 and b_2 = parameters of the equations; R^2 = coefficient of determination; $\bar{d}$ = average diameter (cm); CV% = coefficient of variation in percentage; Trat = Treatment (0%, 40%, 60% and 80% of pruned height).

Among the equations for describing the diameter growth of standing trees from 6 to 15 years old, the five that showed a graphical distribution of residuals that allowed them to be used were: the quadratic and cubic models, the Gompertz, Chapman-Richards and Logistic equations, all with a coefficient of determination (R^2) of around 65%, of which the Chapman-Richards equation showed the lowest coefficient of variation (CV%) and was chosen to describe diameter growth due to its results and ease of interpretation. The results of the selection of diameter growth equations are shown in Appendix II.

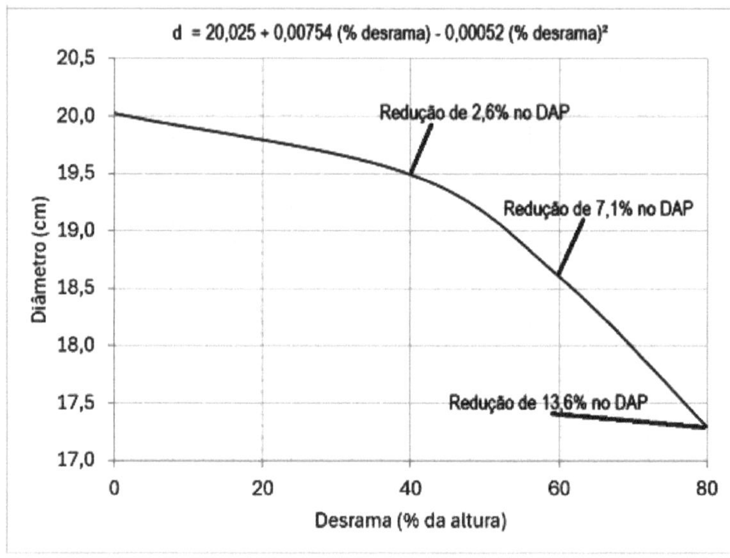

FIGURE 10 - Diameter (cm) estimated as a function of treatment (% of height pruned) at 15 years of age.

The parameters and statistics of the Chapman-Richards equation per treatment, calculated for the period from 6 to 15 years of age, are shown in Table 10. The parameters of a general equation were adjusted for all treatments and the parameters of the equation for each treatment separately. The parameters calculated per treatment were then compared with those found for the general equation.

58

TABLE 10 - Chapman-Richards equation coefficients for d=f(age) by treatment and overall (average) with all treatments, for *P. elliottii*, in Piratini

Treatment	Asymptote A	Growth rate k	Inflection point r	$R^2_{aj.}$	$\bar{y}$	CV%
0	20,17444	0,39406	12,11073	0,70	14,9	19,3
40	20,09953	0,36032	9,64500	0,72	14,6	18,3
60	19,65220	0,28563	5,59595	0,68	13,7	19,3
80	23,17827	0,14210	2,30222	0,64	12,4	21,0
General (average)	19,88003	0,30091	6,37320	0,65	13,9	20,9

Where: *A, k, r* = equation parameters; $\bar{y}$ = mean of the observations; $R^2_{aj.}$ = adjusted coefficient of determination; CV% = coefficient of variation.

Table 10 shows that all the parameters of the 0% and 40% pruning treatments are above the average represented by the general equation with the data from all the treatments, demonstrating that the asymptote, inflection point and growth speed were higher than the average and tended to increase more than the average. The 60% pruning treatment had all the parameters below the average but very close to it. The 80% treatment had only the asymptote above the average and the other parameters below, i.e. the growth speed and the inflection point were lower than the average. In all cases, the standard error of estimate (Syx) remained close to and below 3 cm in diameter and the CV% was between 18.3% and 21%.

The values estimated by the general equation and the individual parameter estimates per treatment of the Chapman-Richards equation are plotted in Figure 11.

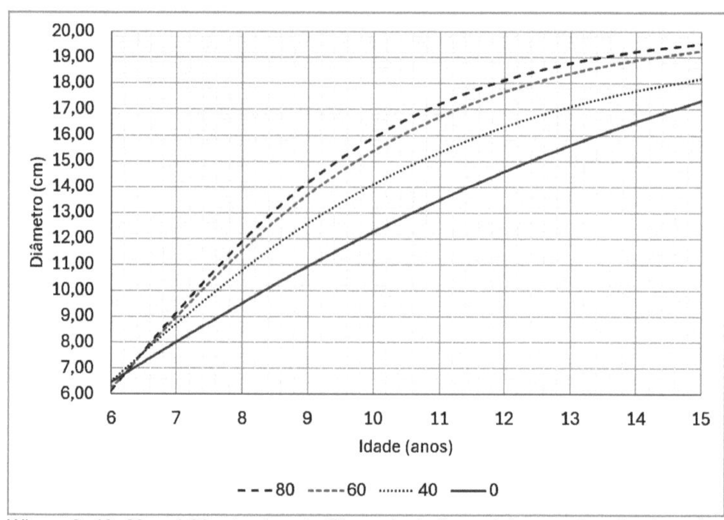

Where: 0, 40, 60 and 80 = treatments (% pruning); General (average)=estimates using the parameters of the Chapman-Richards equation with all treatments (Table 10).

FIGURE 11 - Diameter (cm) estimated by the Chapman-Richards equation per treatment and overall (average) with all treatments, as a function of age (years).

It can be seen that the estimates for the 0% and 40% pruning treatments are above the average, for the 60% treatment the estimates are practically overlapping the average and are below it for the 80% pruning treatment.

The results of the tests to validate the equations in Table 10 are shown in Table 11.

60

TABLE 11 - Validation of Chapman-Richards equations for d=f(age) by treatment

Treatment	White	Durbin-Watson	Kolmogorov-Smirnov
	χ^2	d	D
All	791,06**	1,717^{NS}	0,026**
0	176,63**	1,754 ^{NS}	0,046**
40	185,69**	2,119 ^{NS}	0,033**
60	210,19**	1,989 ^{NS}	0,042**
80	320,02**	2,164 ^{NS}	0,039**

Where: ns=not significant at the 5% probability level; (**) significant at the 5% probability level.

White's test resulted in χ^2 being significant at the 1% probability level, showing a lack of homogeneity of variance; however, it was not taken into account as this was an analysis of a non-linear equation. The Durbin-Watson d value was not significant at the 5% probability level, showing that the residuals are independent. The Kolmogorov-Smirnov D statistic was significant at the 5% probability level, indicating that there is no normality in the distribution of the residuals, and it was not possible to use the t or F tests.

4.1.2 Height (h)

The analysis of variance for each age separately revealed that the influence of the treatments on height growth was minimal, with no significant difference between the average heights of the treatments from 6 to 15 years of age and the small differences that did occur can be considered to be due to chance. The results are shown in Table 12.

TABLE 12 - Evolution of average height (m) with age (years) by pruning treatment and statistical tests, for *P. elliottii*, in Piratini

Age (years)	Pruning treatments				QM_{res}
	0%	40%	60%	80%	
6	3,7	3,9	4,0	4,1	0,190
7	5,3	5,4	5,6	5,5	0,212
8	6,7	6,9	7,0	6,8	0,238
9	8,5	8,6	8,7	8,2	0,167
10	9,8	10,0	10,0	9,5	0,130
11	10,9	10,9	10,7	10,3	0,123
12	12,3	13,4	12,6	11,8	0,352
13	14,1	14,8	14,7	14,0	0,468
14	14,9	15,6	15,4	14,9	0,464
15	16,3	16,9	16,6	16,2	0,432

Where: QM_{res} = mean square of the residual.

Figure 12 shows the evolution of heights by treatment.

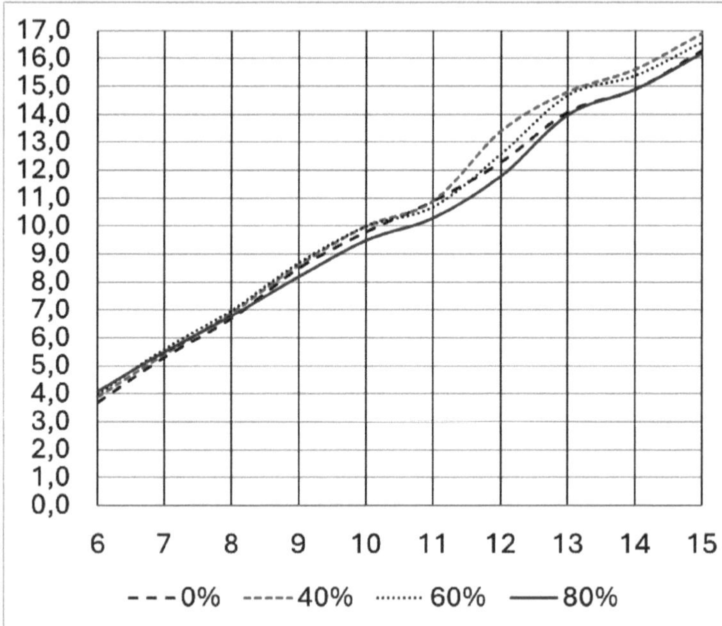

Where: h( )=average height (m); hp( )=pruning height (m); 0, 40, 60 and 80=treatments.
FIGURE 12 - Evolution of average height and pruning height (m) with age (years) per treatment.

It can be seen that the average height curves of the treatments almost overlap, with a slight gap between the ages of 11 and 15.

Figure 12 also shows that the final height of 6 m pruned was reached at different ages depending on the intensity: at 9 years of age in the 40% intensity treatment, at 10 years of age in the 60% intensity treatment and at 14 years of age in the 80% intensity pruning treatment.

At 12 years of age, when there was the greatest differentiation, the treatment with the highest average (40% pruning) had an average height 1.6 m greater than the treatment with the lowest average (80% pruning), but the differences were not significant. At 15 years of age, the averages of the treatments did not differ by more than 0.7 m (4.2%), similar to the standard error of the average, which was 0.66 m; also at this age, the differences were not significant at the 5% probability level by the F test, remaining within the experimental error. The results obtained in this experiment reinforce those of previous experiments. The various studies on pruning conifers cited by Kramer and Kozlowski (1972) show that the operation has no or minimal influence on height growth, as this is processed using the carbohydrates produced in the vicinity of the terminal shoot, and thus does not depend on the branches eliminated by pruning in the lower part of the crown.

4.1.3 Hypsometric ratio

Hypsometric relationship equations were tested for the data collected at 6, 9, 12 and 15 years of age and are shown in Appendix

V. None of the equations tested showed a good fit for all ages in all treatments. However, the use of a joint equation for all treatments resulted in a reasonable fit in almost all cases. At 15 years of age, the models tested had similar results for CV% and adjusted R^2 (Table 13, p. 47) with the data from all treatments.

Considering the adjustment by treatment and age, equations 7, 8 and 9 showed the best results over time in terms of residue distribution, coefficient of variation (CV%) and coefficient of determination (R^2), as shown in Appendix V.

TABLE 13 - Equations for estimating height (m) as a function of diameter (cm) at 15 years of age , for *P. elliottii,* in Piratini

Nº Eq.	Equation	b_0	b_1	b_2	b_3	R^2_{aj}		CV %
1	h=b0+b1.d+b2.d2	5,515	0,9057	-0,0164	-	0,62	16,5	5,3
2	ln h=b0+b1.ln d	1,834	0,3314	-	-	0,65	16,5	5,5
3	ln h= b0+b1(1/d)	3,117	-5,7186	-	-	0,65	16,5	5,3
4	h = b0+b1(1/d)	21,443	-88,8355	-	-	0,63	16,5	5,3
5	h = b0+b1.d+b2(1/d)	21,309	0,0038	-87,724	-	0,62	16,5	5,3
6	h = b0+b1(1/d)+ b2.d2	21,459	-89,014	0	-	0,62	16,5	5,3
7	h=b0+b1.d+ b2(1/d)+b3.d2	16,409	0,29489	-61,714	-0,006	0,62	16,5	5,3
8	h=d2/(b0+b1.d+b2.d2)	2,182	0,08917	0,0488	-	0,65	16,5	5,3
9	h=b0+b1.ln d+b2.ln d2	-21,01	20,8991	-2,7386	-	0,62	16,5	5,3

Where: h=height (m); ln=Neperian logarithm; d=diameter (cm); b , b , b_{012} and b_3 =equation parameters; $\bar{Y}$ =average height (m); CV% = coefficient of variation in percentage.

Among the equations with the best results (7, 8 and 9), the Prodan equation (No. 8) was chosen to describe the hypsometric relationship, as it showed a good distribution of residuals for all treatments at the age of 15 years (Appendix V, p. 89), taken as the basis for comparisons, which was not the case with the others among the best equations.

Between the ages of 6 and 18, the parameters of the Prodan equation, combined with all the treatments, are shown in Table 14, along with the statistical results.

Figure 13 shows the h/d ratio adjusted at ages 9, 12, 15 and 18 using the Prodan equation, highlighting the wide variation at each age. This fact indicates the probable need to estimate an equation per plot in forest inventories and reinforces the need to adjust an equation for each treatment in this experiment.

The covariance test for the Prodan equation, in the model with interaction (Appendix IV, p. 84), resulted in a value of $F = 3.05$ for the interaction, through the sum of squares of type III (sum of squares of corrected products), significant at the 5% level, showing that the slopes are different; the model without interaction, showed $F = 281.16$ for treatments, significant at the 1% level, calculated by the sum of squares of type III, showing that the levels are different.

TABLE 14 - Parameters of the Prodan equation for estimating height for age and statistical results , for *P. elliottii,* in Piratini

Age	b_0	b_1	b_2	$R^2_{aj.}$	$\bar{y}$	CV%
6	-1,12515	1,25166	0,07924	0,76	3,9	9,6
7	0,10899	0,83128	0,08423	0,69	5,5	8,9
8	-9,81678	2,73369	-0,01935	0,24	6,9	11,9
9	3,00449	0,1161	0,08862	0,59	8,5	7,4
10	2,2283	0,17197	0,07812	0,54	9,8	6,9
11	0,90231	0,34175	0,06652	0,39	10,7	8,7
12	-0,53407	0,62362	0,04307	0,56	12,7	0,9
13	-1,26208	0,64362	0,03584	0,65	14,9	7,0
14	0,77361	0,30337	0,04574	0,57	15,2	6,3
15	2,18205	0,08922	0,04885	0,63	16,5	5,3
18	1,47969	0,17075	0,03993	0,54	19,0	6,7

Where: b_0 , b_1 , b_2 = parameters of the equations; $R^2_{aj.}$ = adjusted coefficient of determination; $\bar{y}$ = mean of the observations; CV% = coefficient of variation.

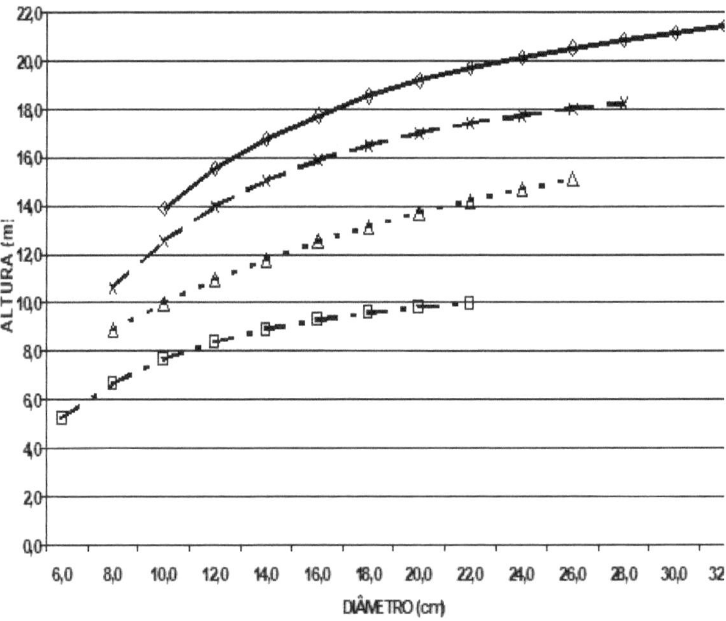

FIGURE 13 - Relationship between diameter (cm) and height (m) at 9, 12, 15 and 18 years using the Prodan equation for all treatments.

Consequently, for greater precision at the age of 15, the parameters of the Prodan equation were adjusted by treatment (Table 15), since at this age the volume was estimated from the diameter and height of the trees in order to compare the average volumetric production.

TABLE 15 - Parameters of the Prodan equation for estimating height per treatment at 15 years of age, for *P. elliottii*, in Piratini

Treatment	b_0	b_1	b_2	$R^2_{aj.}$	$\bar{y}$	CV%
0%	3,43612	0,06110	0,04876	0,64	16,3	5,8
40%	6,02790	-0,41505	0,06393	0,29	16,9	5,2
60%	-0,03707	0,26181	0,04546	0,43	16,6	4,4
80%	0,02646	0,42741	0,03631	0,86	16,2	4,2

Where: Trat = pruning treatments; b , b , b_{012} = equation parameters; $R^2_{aj.}$ = adjusted coefficient of determination; $\bar{y}$ = average of observations; CV% = coefficient of variation.

In the validation of the Prodan equation, at 15 years of age, only the Shapiro-Wilk W value test for the normality of the 40% pruning treatment was significant, indicating that the distribution of the residuals is not normal for that treatment individually. However, this result does not invalidate the use of the function, as it only makes it impossible to use the F and t distributions, which are based on the normal distribution. The other validation statistics were not significant, as shown in Table 16.

TABLE 16 - Validation of Prodan's equations for h=f(d) by treatment and overall with all treatments

Treatment	White χ^2	Durbin-Watson d	Shapiro-Wilk W
0%	3,46 NS	1,611 NS	0,974 NS
40%	4,38 NS	1,698 NS	0,922**
60%	2,82 NS	1,001 NS	0,986 NS
80%	1,14 NS	1,245 NS	0,974 NS
General	1,77 NS	1,243 NS	0,980NS

Where: (NS) not significant at the 5% probability level; (**) significant at the 1% probability level.

4.1.4 Dominant height (h)$_{100}$

As with the average height, there was no significant difference between the averages of the dominant heights (h$_{100}$) of the treatments by the F test, at the 5% probability level, and so Tukey's test was not carried out. The averages and statistical results are shown in Table 17.

TABLE 17 - Evolution of dominant height (m) with age by treatment (m) and statistical tests, for *P. elliottii*, in Piratini

Age (years)	Pruning treatments				QM$_{res}$
	0%	40%	60%	80%	
6	4,7	4,7	4,5	5,0	0,129
7	6,1	6,2	6,1	6,5	0,214
8	7,3	7,9	7,5	7,9	0,230
9	9,2	9,3	9,1	9,2	0,222
10	10,4	10,5	10,5	10,3	0,194
11	11,7	11,3	11,3	11,2	0,180
12	13,3	13,9	13,3	13,0	0,129
13	16,1	15,9	15,3	15,2	0,457
14	16,5	16,5	16,2	16,0	0,209
15	17,4	17,5	17,3	17,7	0,260

Where: QM$_{res}$ = mean square of the residual.

The evolution of the dominant height (h)$_{100}$ is shown in Figure 14, which shows little variation between the treatments. The dominant heights found are comparable to the lowest productivity site found by Tonini (2000) for the region, which showed h_{100} at around 17.5 m.

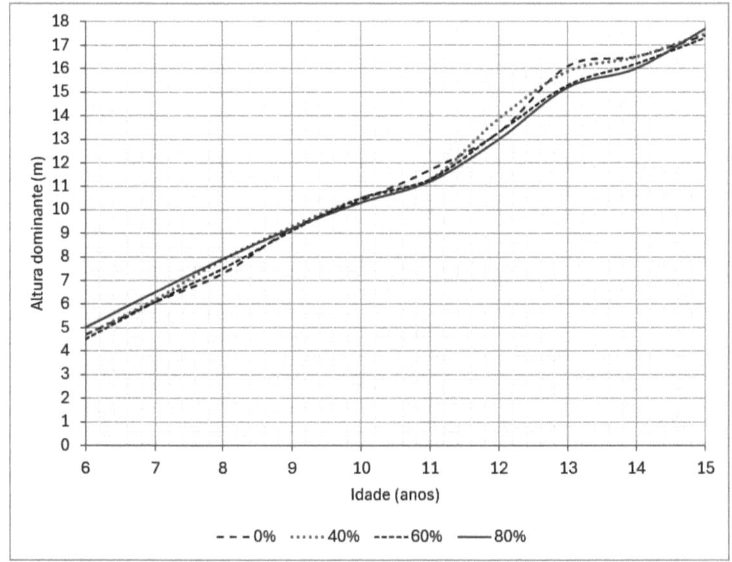

FIGURE 14 - Evolution of dominant height (m) with age by treatment.

4.1.5 Mortality (dead%)

The analysis of variance at the different ages to assess the variation in mortality between pruning treatments revealed a non-significant F (Table 18, p. 52).

Figure 15 shows that mortality increased more in the 0% and 40% pruning treatments over the years. However, the differences were not significant, indicating that the cause cannot be directly attributed to the treatments.

As can be seen, there was practically no increase in mortality between the ages of 7 and 12 and the dead plants from the age of

13 onwards are certainly the result of natural thinning, due to the excessive competition that was established at this age, as shown by the relative spacing index (S%), whose evolution is outlined in Figure 18, and by the stock in basal area (Figure 16, p. 54). This justifies the fact that mortality, from that age onwards, was higher in the 0% and 40% pruning treatments, which showed greater growth, as a result of which competition was established between plants at an earlier age.

TABLE 18 - Cumulative mortality (%) after 6 years, by age and treatment, for *P. elliottii*, in Piratini

Age (years)	Pruning treatments				QM_{res}
	0%	40%	60%	80%	
7	2,0	2,9	1,4	1,4	1,27
8	2,0	2,9	1,4	1,4	1,27
9	2,0	2,9	1,4	1,4	1,27
10	2,0	2,9	1,4	1,8	1,19
11	2,0	2,9	1,4	1,8	1,19
12	2,0	2,9	1,8	1,8	1,19
13	3,8	6,1	1,8	2,9	0,89
14	5,2	7,2	2,5	2,9	0,89
15	7,0	7,2	2,9	3,2	0,85

Where: QM_{res} = mean square of the residual.

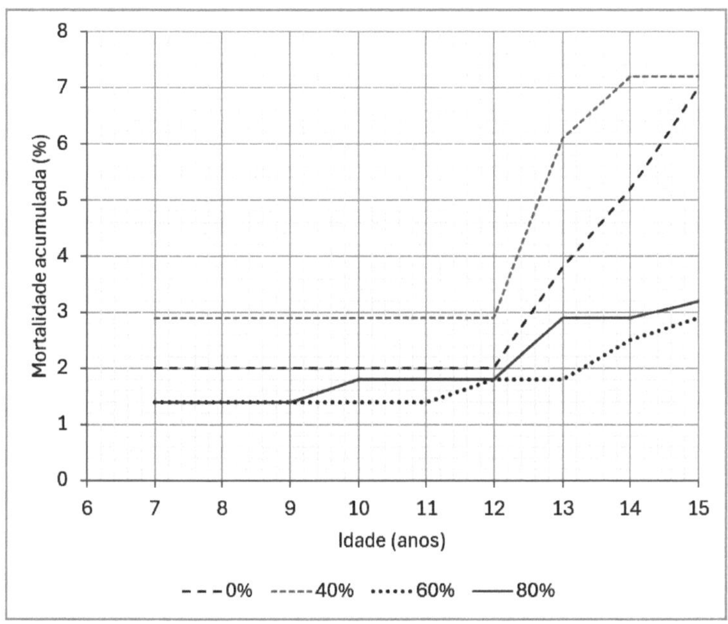

FIGURE 15 - Cumulative mortality (%) after 6 years of age, by age and treatment.

These results are important for the analysis of basal area per hectare in the next section, where it is better characterized that although the differences between the average mortality rates of the treatments are not significant, they can be considered an indirect consequence of the treatments, due to lower growth, delayed competition between plants and postponement of natural thinning in the 60% and 80% pruning intensity treatments.

4.1.6 Basal area per hectare (G)

The stock of basal area per hectare was found to be significantly affected by pruning from the 8th year onwards. The data shows that the differences between the treatments gradually increased from the start of pruning at 6 years of age, when the stocks were similar in all treatments. This occurred until around 10 years of age, when the curves became parallel, even with a slight tendency for the differences to reduce after this age (Table 19 and Figure 16, p. 54).

TABLE 19 - Changes in basal area per hectare (m²/ha) per treatment with age and statistical tests, for *P. elliottii*, in Piratini

Age (years)	Pruning treatments and Tukey's test				QM_{res}
	0%	40%	60%	80%	
6	4,8	5,1	5,5	5,6	3,61
7	10,7	10,5	9,8	8,6	5,91
8	18,0	16,7	15,2	11,5	8,53
9	25,4 a	23,4 a	20,4 a	15,0 b	10,02
10	32,1 a	30,1 a	25,7 a	19,3 b	10,71
11	36,0 a	34,2 a	29,6 a	23,2 b	8,91
12	39,0 a	38,0 a	32,9 a	27,0 b	10,06
13	42,3 a	39,9 a	36,7 a	31,5 b	8,38
14	44,6 a	41,9 a	39,6 a	34,6 b	8,41
15	47,5	45,9	43,1	38,2	9,86

Where: a, b = Tukey's test groups at the 5% probability level; QM_{res} = mean square of the residual.

The evolution of the basal area stock indicates, together with the results of the Tukey test, that the control treatment (0%) and the 40% pruning treatment had the best averages, always accompanied by the 60% treatment in the same group of averages. Only the 80% treatment differed significantly from the others, with the lowest averages.

The basal area stocks at 10 years of age in the 0% and 40% pruning treatments exceed 30 m²/ha, while the 60% and 80% treatments only reach 25.7 and 19.3 m²/ha, respectively.

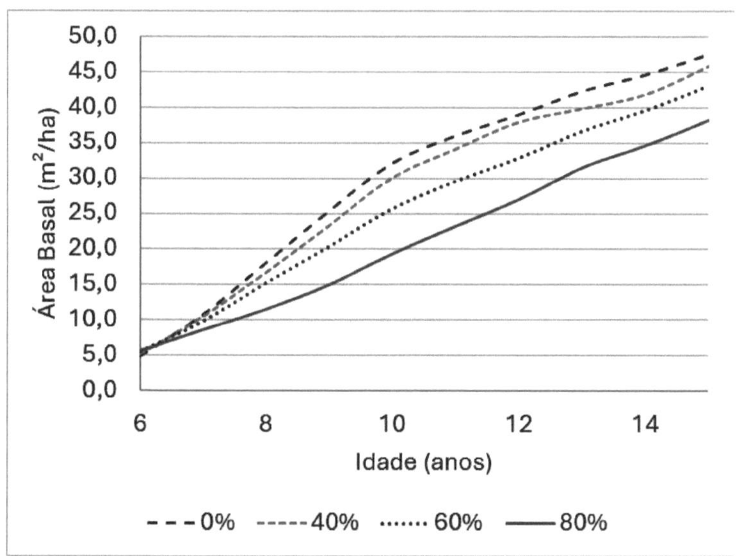

FIGURE 16 - Changes in basal area per hectare (m²/ha) by treatment with age.

It can be seen that, at 9 years of age, the current annual increase in basal area per hectare peaked in the treatment with 0% pruning, which occurred in the 80% treatment only one year later (Figure 17, p. 56).

The slower growth rate caused by pruning therefore delayed the culmination of the increase in basal area per hectare by up to a year and, while competition was established earlier and caused the current annual increase in basal area per hectare (ICA_G) to decrease intensely in the 0% and 40% treatments (Table 20), in the others the increases remained more or less constant.

TABLE 20 - Evolution of the current annual increment in basal area per hectare (m²/ha/year), by treatment, with age and statistical tests, for *P. elliottii*, in Piratini

Age (years)	Pruning treatments and Tukey's test				QM_{res}
	0%	40%	60%	80%	
7	6,0a	5,4a	4,2a	3,0b	1,13
8	7,2a	6,2a	5,4b	2,9c	0,43
9	7,5a	6,7a	5,1b	3,4c	0,30
10	6,6a	6,7a	5,3b	4,3b	0,33
11	3,9	4,1	3,9	3,9	0,52
12	3,0	3,8	3,4	3,8	0,36
13	3,3	1,9	4,0	4,5	1,28
14	2,2	2,0	2,7	3,1	0,34
15	2,8	4,0	3,5	3,6	0,62

Where: iG=annual current increase in basal area per hectare (m²/ha/year); a, b, c = Tukey's test groups at the 5% probability level; QM_{res} = mean square of the residual.

As a result, it can be seen that the current increases of all the treatments practically match each other at 11 years of age (Figure 17, p. 56) and even show reversals of trend after this age, which cannot be directly attributed to the pruning treatments, but is understood to have occurred as an indirect consequence of them.

The sharp drop in increments at 13 and 14 years of age in the 0% and 40% pruning treatments, as shown in Figure 17, is mainly due to increased competition and mortality.

Mortality, observed from 12 to 13 years of age, showed a Pearson correlation coefficient of -59% between the current annual increase in basal area per hectare and the number of dead trees accumulated per treatment, which was significant at the 5% level. This did not occur at the other ages, proving that the drop in increment at that age was related to the death of trees.

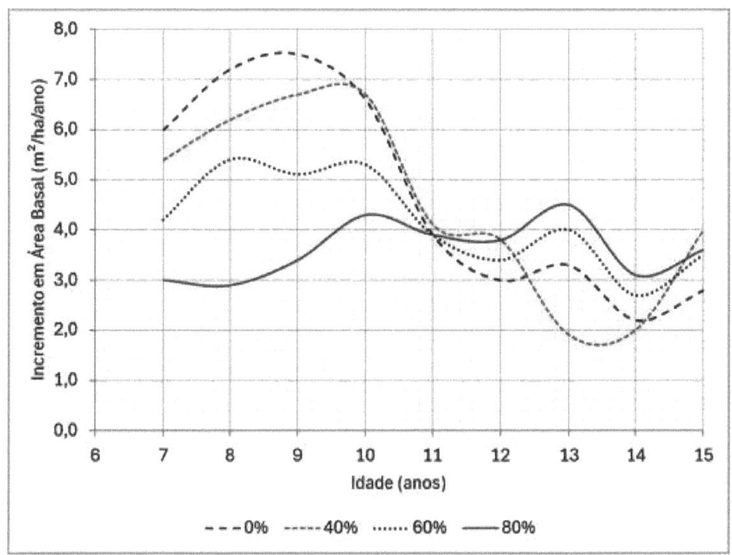

FIGURE 17 - Current annual increase in basal area per hectare (m²/ha/year) by age and treatment.

The process of stand maturation that took place is similar to that described by Oliver *et al.* (2001) and by (Kramer and Kozlowski, 1972): after planting, the trees grew faster and faster and over time competition became established, so growth started to slow down and the less favored trees were suppressed. In the lower-growth treatments, competition was established later and, as a result, natural thinning was also delayed.

4.1.7 Relative spacing index (S%)

Calculation of the relative spacing index revealed that there was excessive competition from the age of 13, when it fell to around 16% (Figure 18, p. 57), the average value found for the region in the Continuous Forest Inventory of Rio Grande do Sul

(UFSM/SEMA-RS, 2003), indicating strong competition between the trees in the stand.

In the analysis of variance for relative spacing (Table 21), using the F test, there was no significant difference at the 5% probability level between the indices of the different treatments.

TABLE 21 - Evolution of the relative spacing index (S%) by treatment with age and statistical tests, for *P. elliottii*, in Piratini

Age (years)	Pruning treatments				QM_{res}
	0%	40%	60%	80%	
6	54	53	55	51	1,25
7	42	41	41	39	0,79
8	35	32	33	32	0,51
9	28	28	28	28	0,21
10	25	24	24	24	0,14
11	22	23	23	23	0,10
12	19	19	19	19	0,10
13	16	16	17	17	0,06
14	16	16	16	16	0,04
15	15	15	15	14	0,04

Where: QM_{res} = mean square of the residual.

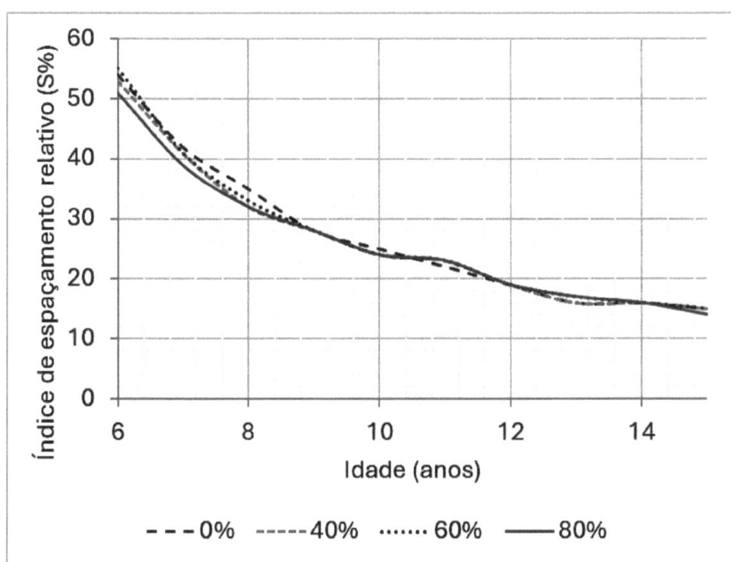

FIGURE 18 - Relative spacing index (S%) by age and treatment.

In Annex VI, the average relative spacing index is graphically plotted as a function of the average individual linear spacing and dominant height at ages 6, 9, 12 and 15. It can be seen that the variation at 12 and 15 years of age in both average spacing and dominant height is small, resulting in indices with a small range of variation.

4.2 Tree volume and shape

The volume and shape study was based on 48 trees (3 per plot) that were felled, cubed and subjected to trunk analysis.

4.2.1 Tree volume

a) Volume modeling

The best volume equation was selected based on the best distribution of residues, the lowest coefficient of variation (CV%) and the highest adjusted R^2. The results of adjusting the volume equations at 15 years of age are shown in Table 22.

TABLE 22 - Test equations for describing the individual volume without bark at 15 years of age, for *P. elliottii*, in Piratini

N° Eq.	b_0	b_1	b_2	b_3	b_4	$R^2_{aj.}$	$\overline{Y}$	CV%
1	-0,171379	0,018591	-	-	-	0,873	0,1782	12,6
2	0,001822	0,000485	-	-	-	0,855	0,1782	13,5
3	0,025978	0,000024	-	-	-	0,871	0,1782	12,7
4	33	0,001288	-0,0000514	0,028546	-	0,911	0,1782	10,8
5	-7,956052	2,110693	-	-	-	0,956	0,1782	13,8
6	-9,690308	0,910790	-	-	-	0,991	0,1782	12,6
7	-10,01054	1,745840	1,102024	-	-	0,992	0,1782	12,6
8	-32,59274	5,241847	-0,605394	13,51356	-2,19195	0,917	0,1782	11,2

Where: b , b , $b_{0\,1\,2}$, b_3 and b_4 = parameters of the equations; $R^2_{aj.}$ = adjusted coefficient of determination; $\overline{y}$ = average individual volume without bark (m³); CV% = coefficient of variation in percentage.

At 15 years of age, the best regression equation for volume was chosen based on the Stoate equation (n°4): $v = b + b_{01}.d + b^2_2 .d^2 .h + b_3 .h$. The Stoate model had the lowest coefficient of variation (10.8%) and an adjusted coefficient of determination of 0.911, which is similar to the other equations, and the residuals are well distributed (Appendix VII, p. 92).

The analysis of covariance for the Stoate equation (Appendix VIII, p. 93) revealed F = 1.10 for the interaction in the corresponding model and F = 0.30 for treatments in the model without the interaction, both non-significant at the 5% probability level, so there was no need to adjust for treatment, as the slopes and levels were similar.

As for the independence of the errors (E_i), for the validation of Stoate's model at 15 years, the values dl=1.40 and du=1.67 were extracted, with d=1.938 > dl=1.40 and d< (4-1.40=2.6); du=1.67 < d=1.938 <4, so d is not significant, H_0 is accepted, there is no serial correlation, the E_i are independent. White's test resulted in a value of 11.24, not significant at 5% probability, concluding that the variance is homogeneous. The Shapiro-Wilk test revealed a value of 0.926, significant at 1%, revealing a lack of normality for the variable analyzed. However, the value is very close to 1 and was not significant when analyzed individually for the 0%, 40% and 60% treatments, and for the 80% treatment it was only significant at the 5% level; this result is considered acceptable. To determine the equations, the diameters of the trees measured standing with bark (d) and the volume without bark (vsc) and oven-dried, of the felled trees were used.

b) Analysis of variance of volume

The results of the analysis of variance on total individual volume are shown in Table 23. The effect of pruning on volume was significant and progressive, since the greater the intensity of pruning, the greater the reduction in volume, as can be seen from the curve shown in Figure 19.

79

TABLE 23 - Analysis of variance of the average individual volume without bark (m³) per plot at 15 years of age per pruning treatment, for *P. elliottii*, in Piratini

FV	GL	SQ	QM	F	R²	$\overline{Y}$	CV %
Blocks	3	0,0002133	0,0000711	1.31ns			
Treatments	3	0,0008247	0,0022749	41,78**	0,9349	0,17371	4,25
Waste	9	0,0004900	0,0000545				
Total	15	0,0075281					

Where: FV=variance factor; GL=degrees of freedom; SQ=sum of squares; QM=mean square; R²=coefficient of determination; $\overline{Y}$ =average of observations; CV%=coefficient of variation in percentage.

According to Table 24, in the treatment with an intensity of 40% pruned height, the reduction observed in the average individual volume was 3.5% in relation to the control treatment without pruning and was not significant at the 5% level using the Tukey test. However, with 60% pruning, the reduction in volume rose to 15.0% and differed from the previous treatments, placing them in group B by Tukey's test. The 80% pruning treatment showed the greatest reduction in volume (26.9%) and was placed in group C by the Tukey test.

TABLE 24 - Tukey's test for average individual bark-free volume (m³) per treatment, at 15 years of age, for *P. elliottii*, in Piratini

Treatment	Average individual volume without bark observed		Group by the Tukey
	(m³)	Reduction compared to control (%)	
0%	0,195585	-	A
40%	0,188955	3,5	A
60%	0,166611	15,0	B
80%	0,143140	26,9	C

Where: Minimum significant difference by Tukey's test= 0.0171 m³.

The diameter was then modeled as a function of pruning intensity, and the F test showed significance at the 1% probability level for the linear (F=98.47) and quadratic (F=17.15) effects. As a result, the equation v=0.196216+0.000220.Trat-0.00001118.Trat² was obtained for the age of 15 years, with R²=0.90 and S_{yx} =0.00766, where v is the individual volume (m³) and Trat is the percentage of height pruned (treatment). The equation is shown in Figure 19, highlighting the downward trend in volume as a function of pruning intensity.

It can be seen that the values estimated by the equation have good precision and are close to the observed values listed in Table 24.

The reductions in volumetric production found by Schneider *et al.* (1999), in a similar experiment with pruning of *P. elliottii,* at 13 years of age and with a maximum pruning height of up to 12 meters, were 12.1% reduction in production in the treatment with 40% pruning and 19.7% in the treatment with 60% of the total height pruned.

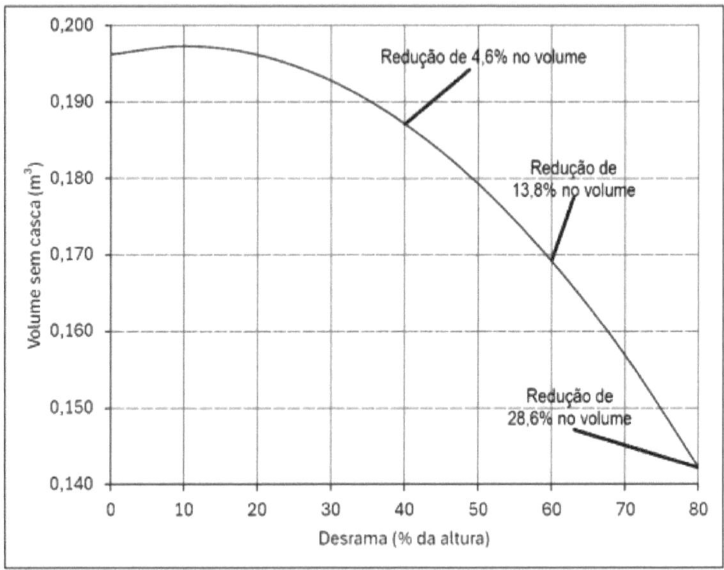

FIGURE 19 - Individual bark-free volume (m³) estimated as a function of treatment (% pruning) at 15 years of age.

Reductions were therefore greater than those obtained in this experiment, but in a stand 2 years younger. The differences between the results of the two experiments can be attributed to two factors: the difference in age of the analyses and the difference in the maximum height of the 12 m and 6 m pruning, since the greater height of the pruning certainly caused an increase in the effect. On the other hand, there is a tendency for the effect of pruning to decrease with age, because as the trees grow in height and emit new branches above the pruned height, the tree crowns recover, restoring their growth capacity; therefore, at the same age, the differences found between the results of the two experiments would possibly be smaller.

4.2.2 Amount of knot-free wood

The results of the analysis of variance on volumes presented in Tables 25 and 26 show that total wood production was significantly higher in the 0% and 40% pruning intensity treatments than in the 60% and 80% treatments. However, the percentage of knot-free wood produced in each treatment in relation to its total volume was not significantly different (Table 26, p. 64). Despite this, the 40% pruning treatment produced around 11% more knot-free volume than the 80% pruning treatment (Table 25, p. 63), which can be explained by the higher total production of the 40% treatment. Although there was no statistical difference at the 5% probability level, this cannot be disregarded in practical terms.

The average percentage of volume produced in the first two logs in relation to the total volume was 62.7%. Of the average volume produced in the first two logs of 0.10425 m³, 73.5% is knot-free wood (0.07658 m³). As there was no significant difference between the treatments in terms of the percentage of knot-free wood produced in each treatment, and considering that the total average individual bark-free volume of the three pruning treatments was 0.16634 m³ per tree felled, with the pruning of the first and second logs, from the age of 6 and up to a height of 6 m, regardless of the treatment, there is an average of close to 46% knot-free wood in each tree.

TABLE 25 - Average individual volume without bark (m³) and average individual volume without knots (m³), at 15 years of age, by pruning treatment, for *P. elliottii*, in Piratini

Treatment	vsc	vsn t_1	vsn t_2	vsn t_1+t_2	vcn t_1+t_2
	(a)	(b)	(c)	(d)	(e)

	m³	(y/y) %	m³	(b/d) %	m³	(c/d) %	m³	(b/d) %	m³	(e/a) %
0	0,1959	100,0	-	-	-	-	-	-	0,1183	60,4
40	0,1892	96,6	0,0530	64,9	0,0287	35,1	0,0817	43,2	0,1082	57,2
60	0,1666	85,0	0,0468	64,0	0,0263	36,0	0,0731	43,9	0,1028	61,7
80	0,1433	73,1	0,0452	60,4	0,0297	39,6	0,0749	52,3	0,1018	71,1
Averages with witness	0,1737		-		-		-		0,1078	62
Averages without witness	0,1663		0,0484		0,0282		0,0766	**46**	0,1043	62,7

Where: vsc = average individual bark-free volume (m³); vsn_{t1} = bark-free and knot-free volume of the 1st log; vsn_{t2} = bark-free and knot-free volume of the 2nd log; vsn_{t1+t2} = sum of the average individual bark-free and knot-free volume of the 1st and 2nd logs; vcn_{t1+t2} = sum of the average individual bark-free and knot-free volume of the 1st and 2nd logs; % = percentage in relation to the column considered (a, b, c, d, e).

Although there were no significant differences, the production of knot-free wood per tree at 15 years of age, shown in Figure 20, was higher in the 40% pruning treatment, as a result of an average total individual volume of 0.18919 m³ and 43.2% knot-free volume, while the 60% pruning treatment had its average total individual volume reduced to 0.16656 m³ and obtained almost the same percentage of knot-free volume, resulting in the lowest knot-free volume production. The 80% pruning treatment, which had its second log pruned earlier than the others and, as a result, obtained a higher percentage of knot-free volume (52.3%), although it had a lower total production per tree (0.14328 m³), obtained a higher knot-free wood production than the 60% pruning treatment, but around 11% lower than the 40% intensity treatment.

TABLE 26 - Results of the analysis of variance on volume at 15 years of age, by pruning treatment, for *P. elliottii*, in Piratini

Response variable	QMres	F for Treatments
Individual total shelled volume (vsc)	0,00005	41,78**
Bark-free and knot-free volume of the 1st log (vsn)$_{t1}$	0,00036	0,56 [NS]
Bark-free and knot-free volume of the 2nd log (vsn)$_{t2}$	0,00013	0,27 [NS]
Bark-free and knot-free volume of the 1st and 2nd logs (vsn)$_{t1+t2}$	0,00089	0,28 [NS]
Volume without bark and with knot of 1st and 2nd logs (vcn $_{t1+t2}$)	0,00126	0,11 [NS]

Where: QMres = mean square of the residual; (NS) not significant at 5% probability level; (**) significant at 1% probability level.

With the results presented, it can be concluded that pruning with an intensity of more than 40% of the total height is not justified, since the amount of wood produced, both in total volume and in knot-free volume, is greater with this intensity of pruning, as can be seen in Figure 20.

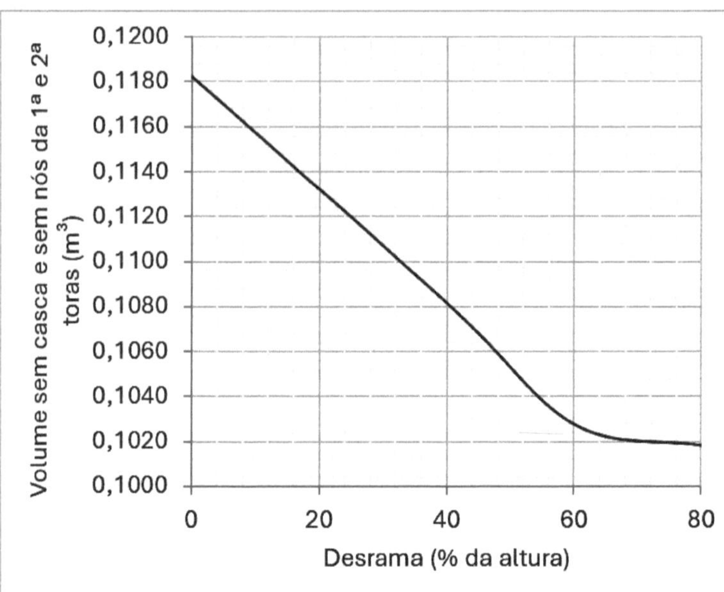

FIGURE 20 - Bark-free and knot-free volume (m³) per tree and per treatment (% pruning) at 15 years of age.

4.2.3 Tree shape

The analysis of variance, presented in Table 27, shows that the artificial form factor ($f = V_r / V_c$), at 15 years of age, showed a significant F at the 5% probability level only for treatments. Tukey's test showed that the 80% (f=0.56776), 40% (f=0.54990) and 60% (f=0.54413) treatments had the highest means and were included in group A, while the 0% treatment (f=0.51332) had the lowest means. The mean of the control treatment was only lower than that of the 80% treatment at the 5% probability level. A correlation test was then carried out and it was found that there was a moderate correlation of 53%, significant at the 1% probability level, between pruning intensity and the artificial form factor (f), indicating that the greater the pruning intensity, the greater the form factor.

TABLE 27 - Analysis of variance for the shape factor (f) of the barkless trunk at 15 years of age, by pruning treatment, for *P. elliottii*, in Piratini

FV	GL	SQ	QM	F	R²	$\bar{Y}$	CV %
Blocks	3	0,00731	0,002435	1,161[NS]	0,294028	0,5438	7,15
Treatments	3	0,01849	0,006163	4,08*			
Waste	41	0,06193	0,001511				
Total	47	0,08773					

Where: FV=variance factor; GL=degrees of freedom; SQ=sum of squares; QM=mean square; R²=coefficient of determination; $\bar{Y}$ =average of observations; CV%=coefficient of variation in percentage; (*) Significant at 5% probability level; (NS) non-significant at 5% probability level.

The shape analysis using the function $g_x = p.x^r$ (Table 28, p. 66), using the parameter "r", showed that the 0% pruning treatment had a value of r=2.1381, showing a conical shape, with a slight tendency towards neiloid. The other pruning treatments, 40% (r=1.7831), 60% (r=1.8371) and 80% (r=1.7869), showed similar values to each other, indicating that the trunks were conical in shape, with a slight paraboloid tendency, differing from the control treatment with no pruning.

Table 28 shows the coefficient values and statistics for the trunk shape equation.

TABLE 28 - Coefficients of the equation $g_x = p.x^r$ at 15 years of age, by pruning treatment, for *P. elliottii*, in Piratini

Treatment	Parameters		Uncorrected Products			Corrected Products	
	p	r	SQ_{Reg}	SQ_{Total}	S_{yx}	SQ_{Total}	R^2
0%	0,00005	2,1381	0,02761	0,02913	0,00039	0,01310	0,8825
40%	0,00012	1,7831	0,02173	0,02309	0,00036	0,01010	0,8655
60%	0,00010	1,8371	0,02038	0,02200	0,00041	0,00953	0,8301
80%	0,00012	1,7869	0,02026	0,02176	0,00039	0,00935	0,8396
General (average)	0,0000949	1,8859	0,08958	0,09598	0,00040	0,042260	0,8481

Where: p, r = parameters of the equation. SQ_{reg} = regression sum of squares. SQ_{tota} l = total sum of squares. S_{yx} = standard error of estimate. CV% = coefficient of variation. R^2 = coefficient of determination.

These results confirm those found through the artificial shape factor (f), in which the control treatment (without pruning) revealed an f of 0.513, lower than the others, showing a less cylindrical shape than the pruning treatments, reinforcing the quotes by Kramer and Kozlowski (1972) about the reduction in trunk taper with pruning intensity.

5 CONCLUSIONS

There was no significant influence of pruning intensity on mortality up to 15 years of age.

The influence of pruning intensity on average height and dominant height was very small and can be disregarded.

Pruning delayed the culmination of ICA in basal area per hectare by up to a year.

The greatest influence of pruning intensity was on tree growth in diameter and volume. There was a reduction in the averages presented by the pruned treatments in relation to the treatment without pruning, at 15 years of age, in the order of:

- 2.0% in diameter and 3.5% in volume in the 40% pruning intensity treatment;

- 7.5% in diameter and 15.0% in volume in the 60% pruning intensity treatment;

- 13.5% in diameter and 26.9% in volume in the 80% pruning intensity treatment.

Each tree pruned to a height of 6 meters from its base produced an average of 46% of its total volume without knot-free bark at 15 years of age, regardless of pruning intensity.

The pruning treatments had a conical shape with a slight paraboloid tendency and the control treatment had a conical shape with a slight neiloid tendency.

6 RECOMMENDATIONS

On experiments with Pinus pruning:
- Include the dry pruning treatment to see if there is a gain in the overall balance, as reported in the literature;
- Measure the height of the first live branch to compare with the height of the pruning and check how much dead branch is being removed in each treatment;
- Measure the diameter of the base of each pruned log until it reaches the maximum diameter for the core;
- If possible, schedule thinning and combine thinning treatments with pruning treatments to enable analysis at a later age;
- Use a greater number of plots to provide a greater number of degrees of freedom in the residual - Gomes (1982) recommends having around 20 degrees of freedom in the residual.

Pruning *Pinus elliottii* Engelm:
- Avoid pruning trees above 40% of their total height as much as possible;
- Only start pruning when the diameter of the base of the logs to be pruned has reached the maximum diameter desired for the rooted core, as pruning is harmful to the trees;
- In commercial pruning, remove all dead or dying branches from the lower part of the crown to avoid the

formation of dead knots, which are more depreciating
and do not contribute to growth.

7 BIBLIOGRAPHICAL REFERENCES

ABREU, Elizabeth C. R.; Scolforo, José R. S.; OLIVEIRA, Antônio D. de O.; MELLO, José M. de; KANEGAE Júnior, Honório. Modeling for early prediction of volume by diametric class for *Eucalyptus grandis*. **Scientia Forestalis**, n. 61, June-2002. p. 86-102.

ANJOS, Adilson dos. **Design of experiments I - Shapiro-Wilk test for normality**. [Curitiba]: UFPR, Department of Statistics, 2003. Available at: <http://www.est.ufpr.br/planexp/planexp.
pdf>. Accessed on: 15/May/2004.

ASSMANN, Ernest. **The principle of forest yield study** [Oxford]: Pergamon press, 1970. 506 p.

BEADLE, Chris; HALL, Malcom. **Pruning - 2** [Hobart]: Private Forests Tasmania, Technical Information Sheet No. 20, Level 2, 2001. Available at: <http://www.privateforests.tas.gov.au/infosheets/20Pruning2.htm>. Accessed on: 21/04/2003.

BUSSAB, Wilton de O. **Analysis of variance and regression.** São Paulo: Atual, 1986. 147 p.

BRITISH COLUMBIA. **Module 7 - Pruning**. Victoria: Province of British Columbia, Ministry of Forests, Forest Practices Branch, 2001. Available at: <http://www.for.gov.bc.ca/hfp/forsite/training/growth-and-yield/toc.htm>. Accessed on: 21/04/2003.

CARVALHO, P.E.R. **Brazilian forest species: silvicultural recommendations, potential and use of wood**. Colombo: EMBRAPA-CNPF / Brasília: EMBRAPA-SPI, 1994. 640 p.

CLEMENTS, Ralph W. **Modern resin methods - manual**. São Paulo: Forestry Institute of the State of São Paulo, 1970. 22 p.

COUTO, Hilton T. Z. do and VETTORAZZO, Sílvia C. Selection of volume and commercial dry weight equations for *Pinus taeda*. **Cerne**, v.5, n.1, 1999. p. 069-080.

EMMINGHAM, Bill and FITZGERALD, Stephen. **Pruning to enhance tree and stand value.** Corvalis: Oregon State University Extension, EC1457, 1995. 12 p.

FINGER, César A. G. **Fundamentals of forest biometrics**. Santa Maria: UFSM, CEPEF-FATEC, 1992. 269 p.

FINGER, C. A. G. and SCHNEIDER, Paulo R. Determination of the thinning weight for *Eucalyptus grandis* Hill ex Maiden forests, based on the relative spacing index. Santa Maria: UFSM/DCF, **Ciência Florestal**, v.9, n.1, 1999. p. 79-87.

FINGER, César A. G.; SCHNEIDER, Paulo R.; BAZZO, José L. ; KLEIN, Jorge E. M. Effect of pruning intensity on the growth and production of *Eucalyptus saligna* Smith. **Cerne**, v.7, n.2, 2001. p. 053-064.

FREUND, Rudolf j.; LITTELL, Ramon C.; SPECTOR, Phillip C. **SAS System for linear models.** Cary: SAS Institute, 1986. 211 p.

GOMES, F. Pimentel. **Curso de estatística experimental**, 10th ed. Piracicaba: Nobel-USP, 1982. 430 p.

GIBSON, Mark D.; CLASON, Terry R.; HILL, Gary L.; GROZDITS, George A. Influence of thinning and pruning on southern pine veneer quality. *In*: **Eleventh Biennial Southern Sivicultural Research Conference**, Department of Agriculture, Forest Service, Southern Research Station, Athens-GA, Nov. 14, 2001. p. 163-167.

GILMAN, Edward F. and WATSON, Dennis G. *Pinus elliottii* - Fact Sheet ST-463. [Gainesville]: United States Forest Service-Environmental Horticulture Department, Florida Cooperative Extension Service, Institute of Food and Agricultural Sciences, University of Florida, Oct. 1994. 4 p.

HIGA, A.R.; MORA, A.L.; STEIN, P. P.; SIMON, A.A.; HIGA, R.C.V. Resistance to rot in *Acacia mearnsii* De Wild. planted in Rio Grande do Sul, Brazil. RESUMEN. *In*: **Primer Congreso Latinoamericano de IUFRO**, Valdivia - Chile, November 22-28, 1998. Available at: <http://iufro.boku.ac.at/iufro/iufronet/d6/wu60304/ ponencias/tema1/higaar.html>. Accessed on: 22/04/2003.

HUSCH, Bertram; MILLER, Charles I.; BEERS, Thomas W. **Forest mensuration**. 3th ed. New York: John Wiley and Sons, 1982. 401 p.

IU. **Testing normality using SAS, STATA, and SPSS**. Indiana: Indiana University - IU, UITS - Center for Statistical and Mathematical Computing. Available at: <http://www.indiana.edu/~statmath>. Accessed on: 15/May/2004. 26 p.

JETER, Jim. Commercial Pruning: An Economic Gamble. **Treasured forests**, Spring 1992. p. 12.

KRAMER, Paul J. and KOZLOWSKI , T. **Fisiologia das árvores**. Lisbon: Calouste Gulbenkian Foundation, 1972. 745 p.

LARSON, Philip R.; KRETSCHMANN, David E.; CLARK, Alexander III; ISEBRANDS, J.G. 2001. **Formation and properties of juvenile wood in**

southern pines: a synopsis. Gen. Tech. Rep. FPL-GTR-129. Madison: U.S. Department of Agriculture, Forest Service, Forest Products Laboratory, Sep/2001. 42 p.

LETSON, Neil. Pruning: A New Look at an Old Practice. **Treasured forests**, Summer 1994. p. 7.

LOETSCH, F.; ZÖHRER, F.; HALLER, K. E. **Forest inventory**, Vol. II. Reinbek: Forest Inventory Section, Federal Research Organization for Forestry and Forest Products, 1973. 469 p.

MACHADO, Sebastião A.; MELLO, José M. de; BARROS Dalmo A. de. Comparison of methods for evaluating total wood volume per unit area for Paraná pine in southern Brazil. **Cerne**, v.6, n.2, 2000. p. 055-066.

MATTOS, João R. de. *Pinus* **species cultivated in Brazil**. São Paulo: Chácaras e Quintais, sd. 133 p.

NCSU. *Pinus elliottii* Engelm - Range and Habitat. Raleigh: North Carolina State University, 2004. Available at: <http://www2.ncsu.edu/unity/locke rs/project/dendrology/index/plantae/vascular/seedplants/gymnosperms/conife rs/pine/pinus/australes/slash/elliottii.html>. Accessed on: 17/07/2004.

NEMEC, Amanda F. Linnell. 1996. **Analysis of repeated measures and time series: an introduction with forestry examples**. Biom. Inf. Handb. 6. Res. Br., B.C. Min. For., Victoria, B.C. Work. Pap. 15/1996. 90 p.

OLIVER, William W.; FERRELL, George T.; TAPPEINER, John C. **Density management of Sierra forests**. Corvallis, Oregon: USDAFS - Pacific Southwest Research Station Redding, California, USDI Forest and Rangeland Ecosystem - Science Center and College of Forestry, Oregon State University, 2001. 7 p.

PERSSON, Reidar and JANZ, Klaus. **Assessment and monitoring of forest and tree resources** - Forest resources assessment. [Rome]: FAO, 2000. p. 16-29.

PRODAN, M. **Mensura forestal**. San José: Inter-American Institute for Cooperation on Agriculture (IICA), 1997. 562 p.

REVISTA DA MADEIRA, Brazilian Association of Wood Producers, Curitiba, Year III, No. 16, 2000.

SAS Institute. **A simple regression model with correction of heteroscedasticity**. Cary: SAS Institute, 2004. Available at: <http:// suport.sas.com/rnd/app/examples/ets/hetero/>. Accessed on: 19/May/2004.

_____. **The SAS System© for Windows© - release 8.02**. Cary: SAS Institute, 2001.

SCHILLING, Ana Cristina. **Influence of pruning on the quality of wood from the first thinning of *Pinus elliottii* Engelm**. Dissertation (Master's Degree in Forestry Engineering) - Master's Degree in Forestry Engineering, UFSM, Santa Maria, 1996. 73 p.

SCHNEIDER, Paulo R. **Regression analysis applied to forestry engineering**. Santa Maria: UFSM, CEPEF, 1998. 236 p.

_____. **Forest management: Planning forest production.** Santa Maria: UFSM, CCR, DCF, CEPEF, 2002. 195 p.

SCHNEIDER, Paulo R.; FINGER, César A. G.; HOPPE, Juarez M. Effect of pruning intensity on the production of *Pinus elliottii* Engelm., planted on poor soil, in the State of Rio Grande do Sul. **Ciência Florestal**, Santa Maria, v.9, n.1, 1999. p. 35-46.

SCHWEINGRUBER, Fritz H. **Tree rings and environment dendroecology**. Vienna: Haupt, 1996. 609 p.

SCOLFORO, José R. S.; RIOS, Múcio S.; OLIVEIRA, Antônio D. de; MELLO, José M. de; MAESTRI, Romualdo. Accuracy of tapering equations to represent the stem profile of *Pinus elliottii*. **Cerne**, v.4, n.1, 1998. p. 100-122.

SCP. **Socio-economic atlas of Rio Grande do Sul**. [Porto Alegre]: SCP-Secretaria da Coordenação e Planejamento do RS, 2004. Available at: [http://www.scp.rs.gov.br]. Accessed on: 25/03/2004.

SEYDACK, A.H.W. **On the limitations of productivity enhancement of tree growth through release thinning in tropical / subtropical forests -** Department of Water Affairs and Forestry - South Africa. Available at: <http://www.dwaf.gov.za/Dir_Forestry/IFM/Docs/CD1/Doc 8 - Seydack.doc>. Accessed on: 21/07/2002.

SIT, Vera. **Catalog of curves for curve fitting**. Victoria, Canada, B.C.: Ministry of Forests, Forest Science Research Branch, Biometrics information handbook series, ISSN 1183-9759-no.4, 1994. 110 p.

SPIECKER, Heinrich. **Forest growth analysis**. Curitiba: Fupef, Technical Series No. 8, 1981. 62 p.

STORCK, Lindolfo; LOPES, Sidinei J. **Experimentação II.** Santa Maria: UFSM, CCR/Dep. Fitotecnia, 1998. 205 p.

STRECK, Edemar; KÄMPF, Nestor; DALMOLI, Ricardo S.D.; KLAMT, Egon; NASCIMENTO, Paulo C. do; SCHNEIDER, Paulo. V.; **Soils of Rio Grande do Sul**. Porto Alegre: EMATER/RS, UFRGS, 2002. 128 p.

TONINI, Helio. **Height growth of *Pinus elliottii* Engelm., in three soil mapping units, in the Serra do Sudeste and Litoral regions, in the State**

of Rio Grande do Sul. Dissertation (Master's Degree in Forest Engineering) - Postgraduate Program in Forest Engineering, Federal University of Santa Maria - RS, Santa Maria, 2000. 113 p.

UFSM/SEMA-RS. Continuous forest inventory - RS. [Santa Maria]: UFSM, SEMA-RS, 2003. Available at: <http://coralx.ufsm.br/ifcrs/>. Accessed on: 07/11/2003.

USDAFS Silvics of North America - *Pinus elliottii* Engelm. Available at: <http://www.cnr.vt.edu/dendro/dendrology/syllabus/pelliottii.htm>. Accessed on: 03/05/2003.

WARNER, Andy. **Pruning**. [Hobart]: Private Forests Tasmania, Technical Information Sheet No. 20, March 1997. Available at: <http://www.privateforests.tas.gov.au/infosheets/20Pruning.htm>. Accessed on: 21/04/2003.

WONNACOTT, Thomas H.; WONNACOTT, Ronald J. **Introduction to statistics**. Rio de Janeiro: Technical and Scientific Books, 1980. 589 p.

ANNEX I - SAS program for analysis of variance of diameters

```
OPTIONS LS=80 PS=40 NODATE NOSTIMER;
DATA MEDICOES;
  INFILE 'C:\DOCUMENTS AND SETTINGS\EDUARDO P. FLORIANO\MEUS
DOCUMENTOS\_DADOS\MEDICOES.DAT';
  INPUT ANO BLOCO TRAT ARVORE DCC H COD @@;
  IDADE=ANO-1985;
PROC SORT; BY IDADE BLOCO TRAT ARVORE;
DATA ANOVAD;
  SET MEDICOES;
  IF COD=0 THEN OUTPUT;
  IF COD>5 AND DCC<9 THEN OUTPUT;
  IF COD>9 THEN DO; COD=0; OUTPUT; END;
PROC SORT DATA=ANOVAD; BY IDADE BLOCO TRAT ARVORE;
PROC GLM DATA=ANOVAD;
  CLASS BLOCO TRAT;
  MODEL DCC=BLOCO TRAT / SS1;
  MEANS BLOCO TRAT / LINES TUKEY;
  BY IDADE;
RUN;
```

ANNEX II - Determination of diameter growth functions

The SAS program for calculating the parameters and statistics
of the equations and a summary of the results are presented below:

PROGRAMA SAS:

```
OPTIONS LS=80 PS=40 NODATE NOSTIMER;
DATA MEDICOES;
   INFILE 'C:\MEUS DOCUMENTOS\_DADOS\MEDICOES.DAT';
   INPUT ANO BLOCO TRAT ARVORE DCC H COD @@;
IF BLOCO=3 AND TRAT=40 AND ARVORE=56 THEN DELETE; /* ESTE EH UM OUTLIER QUE PRECISA
SER ELIMINADO */;
IF COD>0 THEN DELETE;
IDADE=ANO-1985;
D=DCC/10;
T=IDADE;
LNT=LOG(T);
T1=1/T;
T2=T**2;
T3=T**3;
Y=D;
X=T;
IF Y>0 THEN LNY=LOG(Y);
IF IDADE=18 THEN DELETE;
PROC SORT; BY IDADE TRAT ARVORE;
PROC SUMMARY DATA=MEDICOES;
   VAR D;
   BY IDADE TRAT;
   OUTPUT OUT=MEDIAS (KEEP=IDADE TRAT M S) MEAN(D)=M STD(D)=S;
PROC SORT DATA=MEDIAS; BY IDADE TRAT;
DATA DAPCC;
   MERGE MEDIAS MEDICOES; BY IDADE TRAT;
   IF D GT (M+0.75*S) THEN CLASSE=5;
   IF D LE (M+0.75*S) AND D GT (M+0.25*S) THEN CLASSE=4;
   IF D LE (M+0.25*S) AND D GT (M-0.25*S) THEN CLASSE=3;
   IF D LE (M-0.25*S) AND D GT (M-0.75*S) THEN CLASSE=2;
   IF D LE (M-0.75*S) THEN CLASSE=1;
PROC SORT DATA=DAPCC; BY IDADE TRAT ARVORE;
/******************************************************/;
```

```
PROC GLM DATA=DAPCC;
TITLE1 'EQUACOES DE REGRESSAO PARA ANALISE DO CRESCIMENTO EM DAP';
TITLE2 'EQUACAO 1 :   Y = B0 + B1 * T = LINEAR';
  MODEL Y=T / SS1;
  OUTPUT OUT=ARQ1 P=YEST R=RES;
PROC PLOT;PLOT RES*YEST / VREF=0;
/*******************************************************/;
PROC GLM DATA=DAPCC;
TITLE2 'EQUACAO 2 :   Y = B0 + B1 * LOG(T) = LOGARITMICO';
  MODEL Y=LNT / SS1;
  OUTPUT OUT=ARQ2 P=YEST R=RES;
PROC PLOT;PLOT RES*YEST / VREF=0;
/*******************************************************/;
PROC GLM DATA=DAPCC;
TITLE 'EQUACOES DE REGRESSAO PARA ANALISE DO CRESCIMENTO EM DAP';
TITLE2 'EQUACAO 3 :   LOG(Y) = B0 + B1 * LOG(T) - DUPLO LOGARITMICO';
  MODEL LNY=LNT / SS1;
  OUTPUT OUT=ARQ3 P=YEST R=RES;
PROC PLOT;PLOT RES*YEST / VREF=0;
/*******************************************************/;
PROC GLM DATA=DAPCC;
TITLE 'EQUACOES DE REGRESSAO PARA ANALISE DO CRESCIMENTO EM DAP';
TITLE2 'EQUACAO 4 :   Y = B0 + B1 * T**2';
  MODEL Y=T2 / SS1;
  OUTPUT OUT=ARQ4 P=YEST R=RES;
PROC PLOT;PLOT RES*YEST / VREF=0;
/*******************************************************/;
PROC GLM DATA=DAPCC;
TITLE 'EQUACOES DE REGRESSAO PARA ANALISE DO CRESCIMENTO EM DAP';
TITLE2 'EQUACAO 5 :   Y = B0 + B1 * T + B2 * T**2';
  MODEL Y=T T2 / SS1;
  OUTPUT OUT=ARQ5 P=YEST R=RES;
PROC PLOT;PLOT RES*YEST / VREF=0;
/*******************************************************/;
```

```
PROC GLM DATA=DAPCC;
TITLE 'EQUACOES DE REGRESSAO PARA ANALISE DO CRESCIMENTO EM DAP';
TITLE2 'EQUACAO 6 :   Y = B0 + B1 * T + B2 * T**2 + B3 * T**3';
   MODEL Y=T T2 T3 / SS1;
   OUTPUT OUT=ARQ6 P=YEST R=RES;
PROC PLOT;PLOT RES*YEST / VREF=0;
/****************************************************/;
PROC NLIN DATA=DAPCC;
TITLE2 'EQUACAO 7: Y = B0 * T**B1 = POTENCIAL';
PARAMETERS B0=2.0 B1=2.0;
X_B1 = X**B1;
MODEL Y = B0*X_B1;
DER.B0 = X_B1;
DER.B1 = B0*X_B1*LOG(X);
   OUTPUT OUT=ARQ7 P=YEST R=RES;
PROC PLOT;PLOT RES*YEST / VREF=0;
PROC MODEL DATA=DAPCC;
PARAMETERS B0=2.0 B1=2.0;
X_B1 = X**B1;
Y = B0*X_B1;
DER.B0 = X_B1;
DER.B1 = B0*X_B1*LOG(X);
FIT Y;
/****************************************************/;


PROC NLIN DATA=DAPCC;
TITLE2 'EQUACAO 8: Y = EXP( B0 + B1 * T ) = EXPONENCIAL TIPO II OU DE CRESCIMENTO';
PARAMETERS B0=1.8 B1=-0.1;
EABX = EXP(B0 - B1*X);
MODEL Y = EABX;
DER.B0 = EABX;
DER.B1 = -X*EABX;
   OUTPUT OUT=ARQ8 P=YEST R=RES;
PROC PLOT;PLOT RES*YEST / VREF=0;
PROC MODEL DATA=DAPCC;
PARAMETERS B0=1.8 B1=-0.1;
EABX = EXP(B0 - B1*X);
Y = EABX;
DER.B0 = EABX;
DER.B1 = -X*EABX;
FIT Y;
/****************************************************/;
```

```
PROC NLIN DATA=DAPCC;
TITLE2 'EQUACAO 9: Y = B0 * EXP( B1 * T ) = EXPONENCIAL';
PARAMETERS B0=2.0 B1=0.5;
EB1X = EXP(B1*X);
MODEL Y = B0*EB1X;
DER.B0 = EB1X;
DER.B1 = B0*X*EB1X;
  OUTPUT OUT=ARQ9 P=YEST R=RES;
PROC PLOT;PLOT RES*YEST / VREF=0;
PROC MODEL DATA=DAPCC;
PARAMETERS B0=2.0 B1=0.5;
EB1X = EXP(B1*X);
Y = B0*EB1X;
DER.B0 = EB1X;
DER.B1 = B0*X*EB1X;
FIT Y;
/******************************************************/;
```

```
PROC NLIN DATA=DAPCC;
TITLE2 'EQUACAO 10: Y = B0 * {[ 1 - EXP( -B1 * T )]**B2} = CHAPMAN-RICHARDS';
PARAMETERS B0=18.0 B1=0.27 B2=5.0;
EBX = EXP(-B1*X);
EBX1 = 1 - EBX;
EBXB2 = (EBX1)**B2;
MODEL Y = B0*EBXB2;
DER.B0 = EBXB2;
DER.B1 = B0*X*B2*EBX*EBX1**(B2-1);
DER.B2 = B0*EBXB2*LOG(EBX1);
  OUTPUT OUT=ARQ10 P=YEST R=RES;
PROC PLOT;PLOT RES*YEST / VREF=0;
PROC MODEL DATA=DAPCC;
PARAMETERS B0=18.0 B1=0.27 B2=5.0;
EBX = EXP(-B1*X);
EBX1 = 1 - EBX;
EBXB2 = (EBX1)**B2;
Y = B0*EBXB2;
DER.B0 = EBXB2;
DER.B1 = B0*X*B2*EBX*EBX1**(B2-1);
DER.B2 = B0*EBXB2*LOG(EBX1);
FIT Y;
PROC SORT DATA=DAPCC;BY TRAT;
PROC NLIN DATA=DAPCC;
TITLE2 'EQUACAO 10: Y = B0 * {[ 1 - EXP( -B1 * T )]**B2} = CHAPMAN-RICHARDS';
PARAMETERS B0=18.0 B1=0.27 B2=5.0;
EBX = EXP(-B1*X);
EBX1 = 1 - EBX;
EBXB2 = (EBX1)**B2;
MODEL Y = B0*EBXB2;
DER.B0 = EBXB2;
DER.B1 = B0*X*B2*EBX*EBX1**(B2-1);
DER.B2 = B0*EBXB2*LOG(EBX1);
  BY TRAT;
  OUTPUT OUT=ARQ10 P=YEST R=RES;
PROC SORT DATA=ARQ10;BY TRAT;
PROC PLOT;PLOT RES*YEST / VREF=0;
  BY TRAT;
  /******************************************************/;
```

```
PROC NLIN DATA=DAPCC;
TITLE2 'EQUACAO 11: Y = B0/(B3+EXP(B1-B2*X)) = LOGISTICA';
PARAMETERS B0=24.3 B1=3.0 B2=0.5 B3=1.5;
DEN = 1/(B3 + EXP(B1-B2*X));
EE = B0*EXP(B1-B2*X)*DEN**2;
MODEL Y = B0*DEN;
DER.B0 = DEN;
DER.B1 = -EE;
DER.B2 = X*EE;
DER.B3 = -B0*DEN**2;
  OUTPUT OUT=ARQ11 P=YEST R=RES;
PROC PLOT;PLOT RES*YEST / VREF=0;
PROC MODEL DATA=DAPCC;
PARAMETERS B0=24.3 B1=3.0 B2=0.5 B3=1.5;
DEN = 1/(B3 + EXP(B1-B2*X));
EE = B0*EXP(B1-B2*X)*DEN**2;
Y = B0*DEN;
DER.B0 = DEN;
DER.B1 = -EE;
DER.B2 = X*EE;
DER.B3 = -B0*DEN**2;
FIT Y;
/*******************************************************/;


PROC NLIN DATA=DAPCC;
TITLE2 'EQUACAO 12: Y = B0 * EXP[ -EXP( B1 - B2*X ) ] = GOMPERTZ';
PARAMETERS B0=30.0 B1=1.5 B2=0.22;
BDX = EXP(B1-B2*X);
BCBDX = EXP(-BDX);
BBB = BDX*BCBDX*B0;
MODEL Y = BCBDX*B0;
DER.B0 = BCBDX;
DER.B1 = -BBB;
DER.B2 = X*BBB;
  OUTPUT OUT=ARQ12A P=YEST R=RES;
PROC PLOT DATA=ARQ12A;
  PLOT RES*YEST / VREF=0;
PROC MODEL DATA=DAPCC;
PARAMETERS B0=30.0 B1=1.5 B2=0.22;
BDX = EXP(B1-B2*X);
BCBDX = EXP(-BDX);
BBB = BDX*BCBDX*B0;
Y = BCBDX*B0;
DER.B0 = BCBDX;
DER.B1 = -BBB;
DER.B2 = X*BBB;
FIT Y;
RUN;
```

Equation	b0	b1	b2	b3	R²aj.	$\bar{y}$	CV%
1	-0,23128	1,35146	-	-	0,62	13,9	22,0
2	-17,6811	13,69027	-	-	0,64	13,9	21,2
3	-0,14582	1,17057	-	-	0,75	13,9	22,6
4	6,64185	0,06173	-	-	0,57	13,9	23,2
5	-13,07875	4,00252	-0,1265	-	0,65	13,9	21,0
6	-23,91208	7,41871	-0,46735	0,01083	0,65	13,9	20,9
7	1,39071	0,98207	-	-	0,62	13,9	22,0
8	1,65509	-0,09048	-	-	0,58	13,9	23,2
9	5,23592	0,09044	-	-	0,58	13,9	23,2
10	19,87287	0,3012	6,38545	-	0,65	13,9	20,9
11	28,45793	3,73238	0,44865	1,5	0,65	13,9	21,0
12	19,69056	2,05565	0,32206	-	0,65	13,9	21,0

Where: b0, b1, b2, b3=equation parameters; N=number of observations; K=number of equation parameters; R²aj.=adjusted coefficient of determination; $\bar{y}$ =average of observations; CV%=coefficient of variation in percentage.

The SAS program and the graph produced of diameter (cm) as a function of pruning intensity (%) over time (years) are shown below.

SAS program:

```
OPTIONS LS=70 PS=60 NODATE NOSTIMER;
DATA A;
 DO TRAT=0 TO 80 BY 1;
  DO IDADE=6 TO 15 BY 1;
   IF IDADE=6  THEN D=5.925+0.00263*TRAT+0.000042*TRAT**2;
   IF IDADE=7  THEN  D=9.185+0.00595*TRAT-0.000234*TRAT**2;
   IF IDADE=8  THEN D=11.976+0.01331*TRAT-0.000546*TRAT**2;
   IF IDADE=9  THEN D=14.307+0.01519*TRAT-0.000727*TRAT**2;
   IF IDADE=10 THEN D=16.087+0.01800*TRAT-0.000815*TRAT**2;
   IF IDADE=11 THEN D=17.023+0.01851*TRAT-0.000787*TRAT**2;
   IF IDADE=12 THEN D=17.701+0.02232*TRAT-0.000782*TRAT**2;
   IF IDADE=13 THEN D=18.583+0.01085*TRAT-0.000587*TRAT**2;
   IF IDADE=14 THEN D=19.233+0.00715*TRAT-0.000526*TRAT**2;
   IF IDADE=15 THEN D=20.025+0.00754*TRAT-0.000526*TRAT**2;
   OUTPUT;
   END;
 END;
PROC PRINT;
PROC G3D;
  LABEL TRAT='DESRAMA (%)';
  LABEL IDADE='IDADE (ANOS)';
  LABEL D='d (CM)';
  PLOT IDADE*TRAT=D / GRID
          XTICKNUM=9 YTICKNUM=10 ZTICKNUM=9
          ZMAX=21 ZMIN=5
          ROTATE=30;
 RUN;
```

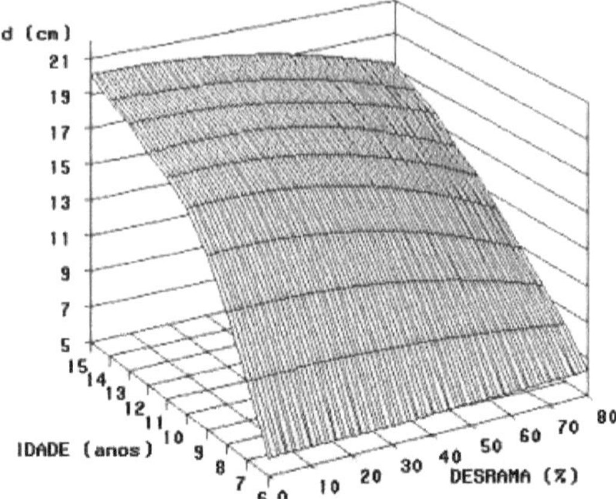

Where: d = diameter (cm); Pruning = pruning height as a percentage of total tree height.

Graph 2 - Diameter (cm) as a function of pruning intensity (%) over time (years).

ANNEX IV - Analysis of covariance for hypsometric ratio

The SAS program and the results of the analysis of covariance
for the hypsometric ratio are presented below:

```
OPTIONS LS=80 PS=40 NODATE NOSTIMER;
TITLE1 'RELACAO HIPSOMETRICA - TESTE DE COVARIANCIA - 2000';
DATA MEDICOES;
  INFILE 'C:\DOCUMENTS AND SETTINGS\EDUARDO P. FLORIANO\MEUS
DOCUMENTOS\_DADOS\MEDICOES.DAT';
  INPUT ANO BLOCO TRAT ARVORE D H COD @@;
  IDADE=ANO-1985;
  IF COD=0 AND H>9 AND IDADE=15 THEN OUTPUT;
PROC SORT; BY IDADE BLOCO TRAT ARVORE;
DATA ALTURA;
  SET MEDICOES;
  D=D/10;
  /* EQUACAO DE PRODAN CONJUNTA PARA OS 4 TRATS */;
  YEST=D**2/(2.18205+0.08922*D+0.04885*D**2);
PROC SORT DATA=ALTURA; BY IDADE TRAT;
/********* TESTE DA DIFERENÇA ENTRE NIVEIS *********/;
PROC GLM DATA=ALTURA;
  CLASS TRAT;
  MODEL H = YEST TRAT / SOLUTION SS1 SS3;
  LSMEANS TRAT / E STDERR PDIFF;
/******* TESTE DA DIFERENÇA ENTRE INCLINACOES ******/;
PROC GLM DATA=ALTURA;
  CLASS TRAT;
  MODEL H = YEST TRAT TRAT*YEST / SOLUTION SS1 SS3;
  LSMEANS TRAT / E STDERR PDIFF;
RUN;
```

Difference between levels:

```
                           THE SAS SYSTEM

            HYPSOMETRIC RELATIONSHIP - COVARIANCE TEST - 2000
                    GENERAL LINEAR MODELS PROCEDURE
                       CLASS LEVEL INFORMATION
                        CLASS LEVELS VALUES
                         TRAT 4 0 40 60 80
                 NUMBER OF OBSERVATIONS IN DATA SET = 159

                         THE GLM PROCEDURE
DEPENDENT VARIABLE: H
                                      SUM OF
 SOURCE DF SQUARES MEAN SQUARE F VALUE PR > F
 MODEL 4 21765.27906 5441.31976 77.80 <.0001
 ERROR 154 10771.13604 69.94244
 CORRECTED TOTAL 158 32536.41509

             R-SQUARE COEFF VAR ROOT MSE H MEAN
             0.668951 5.073803 8.363160 164.8302


 SOURCE DF TYPE I SS MEAN SQUARE F VALUE PR > F
 YEST 1 20491.68890 20491.68890 292.98 <.0001
 TRAT 3 1273.59016 424.53005 6.07 0.0006

 SOURCE DF TYPE III SS MEAN SQUARE F VALUE PR > F
 YEST 1 19664.82486 19664.82486 281.16ᵃ <.0001
 TRAT 3 1273.59016 424.53005 6.07 0.0006
                                      STANDARD
     PARAMETER ESTIMATE ERROR T VALUE PR > |T|
     INTERCEPT -7.64460033 B 10.04520037 -0.76 0.4478
     YEST 10.55902767 0.62972260 16.77 <.0001
     TRAT 0 -6.52600766 B 1.99136606 -3.28 0.0013
     TRAT 40 -0.44773647 B 1.96832151 -0.23 0.8204
     TRAT 60 0.73578435 B 1.92186648 0.38 0.7024
     TRAT 80 0.00000000 B . . .
NOTE: THE X'X MATRIX HAS BEEN FOUND TO BE SINGULAR, AND A GENERALIZED INVERSE
      WAS USED TO SOLVE THE NORMAL EQUATIONS.  TERMS WHOSE ESTIMATES ARE
      FOLLOWED BY THE LETTER 'B' ARE NOT UNIQUELY ESTIMABLE.

                         LEAST SQUARES MEANS
                 COEFFICIENTS FOR TRAT LEAST SQUARE MEANS
                                TRAT LEVEL
 EFFECT 0 40 60
 INTERCEPT 1 1 1
 YEST 16.4834036 16.4834036 16.4834036
 TREAT 0 1 0 0
 TRAT 40 0 1 0
 TRAT 60 0 0 1
 TREAT 80 0 0 0
```

ᵃ F for testing the difference between levels.

```
                    COEFFICIENTS FOR TRAT LEAST SQUARE MEANS
                                                 TRAT LEVEL
               EFFECT 80
               INTERCEPT 1
               YEST 16.4834036
               TREAT 0 0
               TRAT 40 0
               TRAT 60 0
               TRAT 80 1
                                        STANDARD LSMEAN
         TRAT H LSMEAN ERROR PR > |T| NUMBER
         0 159.878107 1.348081 <.0001 1
         40 165.956378 1.336191 <.0001 2
         60 167.139899 1.339579 <.0001 3
         80 166.404114 1.388093 <.0001 4


                         LEAST SQUARES MEANS
                  LEAST SQUARES MEANS FOR EFFECT TRAT
                  PR > |T| FOR HO: LSMEAN(I)=LSMEAN(J)
                       DEPENDENT VARIABLE: H
         I/J 1 2 3 4
            1 0.0014 0.0002 0.0013
            2 0.0014 0.5333 0.8204
            3 0.0002 0.5333 0.7024
            4 0.0013 0.8204 0.7024
NOTE: TO ENSURE OVERALL PROTECTION LEVEL, ONLY PROBABILITIES ASSOCIATED WITH
      PRE-PLANNED COMPARISONS SHOULD BE USED.
```

Difference between inclinations:

```
                         THE SAS SYSTEM

         HYPSOMETRIC RELATIONSHIP - COVARIANCE TEST - 2000
                  GENERAL LINEAR MODELS PROCEDURE

                     CLASS LEVEL INFORMATION
                     CLASS LEVELS VALUES
                     TRAT 4 0 40 60 80
              NUMBER OF OBSERVATIONS IN DATA SET = 159

DEPENDENT VARIABLE: H
                                 SUM OF
 SOURCE DF SQUARES MEAN SQUARE F VALUE PR > F
 MODEL 7 22380.19062 3197.17009 47.53 <.0001
 ERROR 151 10156.22447 67.25976
 CORRECTED TOTAL 158 32536.41509

              R-SQUARE COEFF VAR ROOT MSE H MEAN
              0.687851 4.975548 8.201205 164.8302


 SOURCE DF TYPE I SS MEAN SQUARE F VALUE PR > F
 YEST 1 20491.68890 20491.68890 304.66 <.0001
 TRAT 3 1273.59016 424.53005 6.31 0.0005
 YEST*TRAT 3 614.91157 204.97052 3.05 0.0306

 SOURCE DF TYPE III SS MEAN SQUARE F VALUE PR > F
 YEST 1 13252.20382 13252.20382 197.03 <.0001
```

TRAT 3 678.66101 226.22034 3.36 0.0204
YEST*TRAT 3 614.91157 204.97052 3.05ᵃ 0.0306
                                        STANDARD
     PARAMETER ESTIMATE ERROR T VALUE PR > |T|
     INTERCEPT -21.13783666 B 14.27880750 -1.48 0.1409
     YEST 11.41232944 B 0.89924940 12.69 <.0001
     TRAT 0 -22.71940218 B 25.59733493 -0.89 0.3762
     TRAT 40 62.63945269 B 31.86023832 1.97 0.0511
     TRAT 60 60.88157635 B 29.60741198 2.06 0.0415
     TRAT 80 0.00000000 B . . .
     YEST*TRAT 0 0.90332290 B 1.54371653 0.59 0.5593
     YEST*TRAT 40 -3.80739320 B 1.91854951 -1.98 0.0490
     YEST*TRAT 60 -3.69255086 B 1.81491348 -2.03 0.0436
     YEST*TRAT 80 0.00000000 B . . .
NOTE: THE X'X MATRIX HAS BEEN FOUND TO BE SINGULAR, AND A GENERALIZED INVERSE
      WAS USED TO SOLVE THE NORMAL EQUATIONS. TERMS WHOSE ESTIMATES ARE
      FOLLOWED BY THE LETTER 'B' ARE NOT UNIQUELY ESTIMABLE.

                        LEAST SQUARES MEANS
             COEFFICIENTS FOR TRAT LEAST SQUARE MEANS
                           TRAT LEVEL
EFFECT 0 40 60
INTERCEPT 1 1 1
YEST 16.4834036 16.4834036 16.4834036
TREAT 0 1 0 0
TRAT 40 0 1 0
TRAT 60 0 0 1
TREAT 80 0 0 0
YEST*TRAT 0 16.4834036 0 0
YEST*TRAT 40 0 16.4834036 0
YEST*TRAT 60 0 0 16.4834036
YEST*TRAT 80 0 0 0

              COEFFICIENTS FOR TRAT LEAST SQUARE MEANS
                                      TRAT LEVEL
             EFFECT 80
             INTERCEPT 1
             YEST 16.4834036
             TREAT 0 0
             TRAT 40 0
             TRAT 60 0
             TRAT 80 1
             YEST*TRAT 0 0
             YEST*TRAT 40 0
             YEST*TRAT 60 0
             YEST*TRAT 80 16.4834036

                        LEAST SQUARES MEANS
                          STANDARD LSMEAN
         TRAT H LSMEAN ERROR PR > |T| NUMBER
         0 159.146629 1.398032 <.0001 1
         40 166.856849 1.395839 <.0001 2
         60 166.991966 1.315810 <.0001 3
         80 166.976195 1.430023 <.0001 4

ᵃ F for testing the difference between slopes.

110

```
                    LEAST SQUARES MEANS FOR EFFECT TRAT
                    PR > |T| FOR HO: LSMEAN(I)=LSMEAN(J)
                           DEPENDENT VARIABLE: H
          I/J 1 2 3 4
             1 0.0001 <.0001 0.0001
             2 0.0001 0.9439 0.9525
             3 <.0001 0.9439 0.9935
             4 0.0001 0.9525 0.9935
NOTE: TO ENSURE OVERALL PROTECTION LEVEL, ONLY PROBABILITIES ASSOCIATED WITH
      PRE-PLANNED COMPARISONS SHOULD BE USED.
```

ANNEX V - Distribution of residuals, CV% and R²(%) for the hypsometric relationship equations tested at 6, 9, 12 and 15 years of age

Statistics	No. eq	0% Age (years)				40% Age (years)				60% Age (years)				80% Age (years)				Pre-
		6	9	12	15	6	9	12	15	6	9	12	15	6	9	12	15	
Waste distribution	1	R	P	P	R	B	P	P	R	R	B	P	R	B	P	R	P	7
	2	P	R	P	B	P	P	R	B	B	B	R	R	P	P	B	R	5
	3	P	P	P	R	P	P	R	R	P	P	R	R	P	R	P	P	9
	4	P	P	P	B	P	P	R	B	P	B	P	B	P	P	P	P	8
	5	P	R	B	B	R	P	R	R	B	B	B	R	R	P	B	P	4
	6	P	P	R	B	R	P	R	R	B	B	R	R	R	P	R	P	6
	7	B	R	O	B	O	P	R	R	B	R	B	O	R	P	O	R	1
	8	P	R	B	B	O	P	P	B	B	B	B	B	P	P	B	B	3
	9	B	R	B	B	O	P	P	R	B	R	B	R	O	P	B	R	2
CV%	1	10	8	10	6	10	6	10	5	8	6	5	4	9	6	7	4	7
	2	10	8	10	6	10	6	10	5	8	6	5	4	9	7	7	4	5
	3	14	8	11	6	12	6	10	5	9	6	5	4	13	7	7	4	2
	4	15	8	11	6	13	7	10	5	10	6	5	4	14	7	7	5	1
	5	10	8	10	6	10	6	10	5	8	6	5	4	9	6	7	4	7
	6	11	8	11	6	10	6	10	5	8	6	5	4	9	7	7	4	3
	7	10	8	10	6	10	6	10	5	8	6	5	4	9	6	6	4	9
	8	10	8	10	6	10	8	10	5	8	6	5	4	9	6	7	4	4
	9	10	8	10	6	10	6	10	5	8	6	4	4	9	6	7	4	9
R² aj.(%)	1	77	59	61	66	77	71	18	31	82	62	67	45	79	76	74	85	6
	2	76	65	58	67	74	75	18	30	80	65	64	45	71	81	74	90	8
	3	26	60	46	67	50	59	18	31	63	63	58	44	31	68	66	83	1
	4	40	58	53	66	60	64	19	31	70	62	61	45	44	70	69	84	2
	5	76	59	61	66	77	71	18	32	82	62	68	44	79	75	73	86	5
	6	73	59	60	66	76	70	18	32	71	62	68	44	79	75	73	86	4
	7	78	59	63	66	77	71	18	32	82	62	67	45	79	75	75	86	9
	8	74	59	61	66	77	71	18	32	82	62	68	44	40	75	74	86	3
	9	77	59	61	66	77	71	18	32	82	62	68	44	79	76	74	86	7

Where: 0%, 40%, 60%, 80% = treatments; No. eq. = equation number; Distribution of residues: O = optimal, B = good, R = reasonable, P = poor; CV% = coefficient of variation in percent; R²aj. (%) = adjusted coefficient of determination in percent.

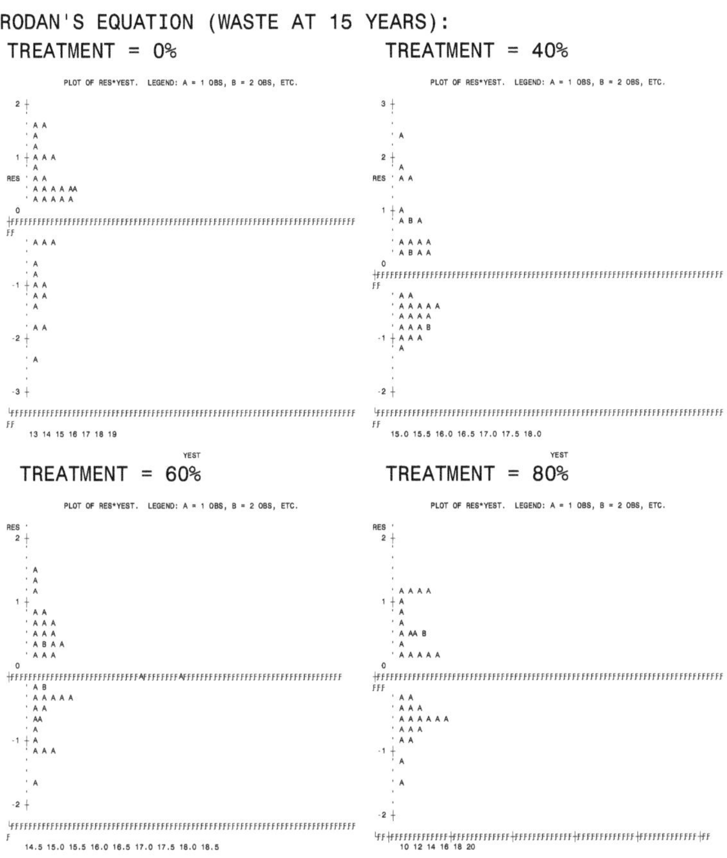

Graph 3 - Residuals of the Prodan equation by treatment at 15 years of age for h=f(d).

ANNEX VI - Relative spacing index (S%) at 6, 9, 12 and 15 years of age

IDADE=6

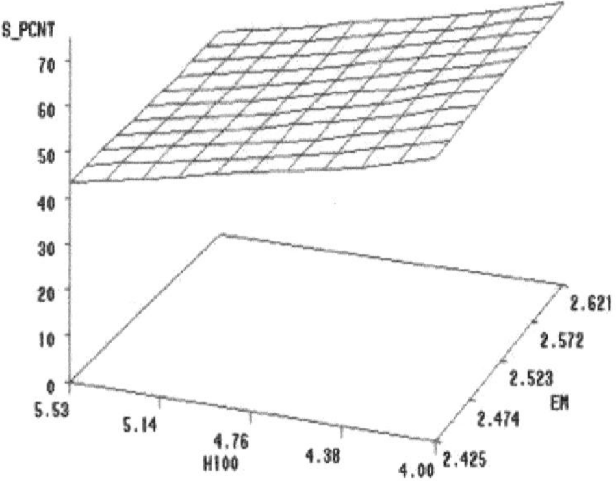

IDADE=9

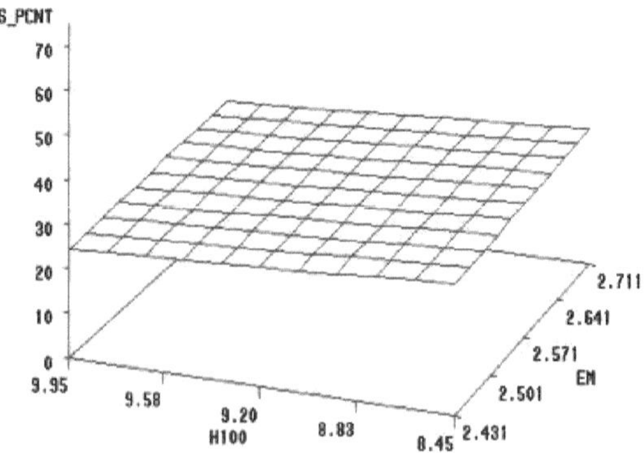

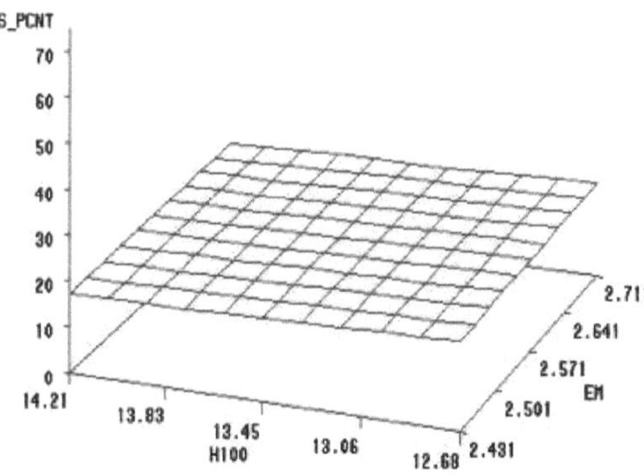

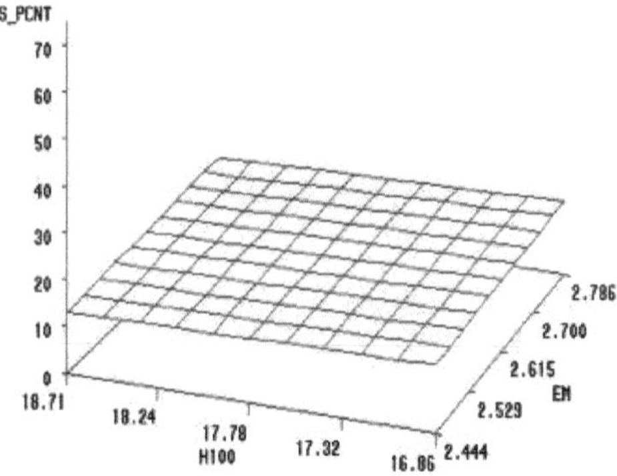

Where: S_PCNT= relative spacing index (S%); H100= dominant height (m); MS = average linear spacing (m).

Graph 4 - Relative spacing index (S%) at 6, 9, 12 and 15 years.

115

EQUATION 4 : V = B0 + B1 * D**2 + B2 * D**2*H + B3 * H

WASTE

PLOT OF RES*YEST. LEGEND: A = 1 OBS, B = 2 OBS, ETC.

```
 RES '
 0.04 ^
      '
      '
    ' A
    ' A A
 0.02 ^ A
    ' A A A
    ' A A A A
    ' A A A
    ' B A A AA
 0.00 ^FFFFFFFFFFFFAFFAAAAFFFFFFFFFFFFFFAFFFFFFFFFFFAFFFFAFFFFFFFFFAAFFFFFFFFFFFFFFFFFF
    ' A A A B
    ' A A A A
    ' A A A
      '
-0.02 ^ A
    ' A
      '
      '
      '
-0.04 ^ A
      '
      '
      '
    ' A
-0.06 ^
      '

 $FFFFFFFFFFFFFFFFFFFFFFFFFFFFFFFFFFFFFFFFFFFFFFFFFFFFFFFFFFFFFFFFFFFFFFFFFFFFFFFFFFF
 FF
    0.05 0.10 0.15 0.20 0.25 0.30
```

ESTIMATED VOLUME (M³)

Graph 5 - Residuals for the Stoate equation at 15 years.

ANNEX VIII - Analysis of covariance for the Stoate equation

Difference between levels:

```
                THE SAS SYSTEM
                THE GLM PROCEDURE
DEPENDENT VARIABLE: V
                SUM OF
SOURCE          DF    SQUARES   MEAN SQUARE  F VALUE  PR > F
MODEL           4   0.10230155  0.02557539   13.32  <.0001
ERROR          43   0.08254127  0.00191956
CORRECTED TOTAL     47   0.18484282


      R-SQUARE   COEFF VAR    ROOT MSE     V MEAN
      0.553452   24.59084    0.043813    0.178167


SOURCE          DF  TYPE III SS   MEAN SQUARE   F VALUE  PR > F
YEST            1   0.09896563   0.09896563    51.56  <.0001
       TRAT              3    0.00172614    0.00057538    0.30  0.8254
```

Difference between inclinations:

```
                                  THE  SAS  SYSTEM
                THE GLM PROCEDURE
DEPENDENT VARIABLE: V
                SUM OF
SOURCE          DF    SQUARES   MEAN SQUARE  F VALUE  PR > F
MODEL           7   0.10857581  0.01551083    8.14  <.0001
ERROR          40   0.07626701  0.00190668
CORRECTED TOTAL     47   0.18484282


      R-SQUARE   COEFF VAR    ROOT MSE     V MEAN
      0.587395   24.50815    0.043665    0.178167


SOURCE          DF  TYPE III SS  MEAN SQUARE  F VALUE  PR > F
YEST            1   0.10053814   0.10053814   52.73  <.0001
TRAT            3   0.00223266   0.00074422    0.39  0.7606
       yest*TRAT          3    0.00627426    0.00209142    1.10  0.3616
```

yes

I want morebooks!

Buy your books fast and straightforward online - at one of world's fastest growing online book stores! Environmentally sound due to Print-on-Demand technologies.

Buy your books online at
www.morebooks.shop

Kaufen Sie Ihre Bücher schnell und unkompliziert online – auf einer der am schnellsten wachsenden Buchhandelsplattformen weltweit! Dank Print-On-Demand umwelt- und ressourcenschonend produzi ert.

Bücher schneller online kaufen
www.morebooks.shop

Printed by Books on Demand GmbH, Norderstedt / Germany